# Advances in the Studies of the Benthic Zone

*Edited by Luis A. Soto*

Published in London, United Kingdom

IntechOpen

*Supporting open minds since 2005*

Advances in the Studies of the Benthic Zone
http://dx.doi.org/10.5772/intechopen.81961
Edited by Luis A. Soto

Contributors
María Custodio, Richard Peñaloza, Heidi De La Cruz, Ana Maria Pires-Vanin, Giovanni Chimienti, Francesco Mastrototaro, Gianfranco D'Onghia, María A. Mendoza-Becerril, José Agüero, Crisalejandra Rivera, Steinar Daae Johansen, Luis Soto, Diego Lopez-Veneroni, Åse Emblem



Notice
Statements and opinions expressed in the chapters are these of the individual contributors and not necessarily those of the editors or publisher. No responsibility is accepted for the accuracy of information contained in the published chapters. The publisher assumes no responsibility for any damage or injury to persons or property arising out of the use of any materials, instructions, methods or ideas contained in the book.

First published in London, United Kingdom, 2020 by IntechOpen
IntechOpen is the global imprint of INTECHOPEN LIMITED, registered in England and Wales, registration number: 11086078, 7th floor, 10 Lower Thames Street, London, EC3R 6AF, United Kingdom
Printed in Croatia

British Library Cataloguing-in-Publication Data
A catalogue record for this book is available from the British Library

Additional hard and PDF copies can be obtained from orders@intechopen.com

Advances in the Studies of the Benthic Zone
Edited by Luis A. Soto
p. cm.
Print ISBN 978-1-83880-043-7
Online ISBN 978-1-83880-044-4
eBook (PDF) ISBN 978-1-83880-989-8

# Meet the editor

 Luis A. Soto, BS, Diploma FAO-VNIRO, MSc, PhD, is a biological oceanographer at the Institute of Marine Sciences and Limnology, UNAM, Mexico (2019). He was Dean of Graduate Students in the Marine Science Program (1983–1987) and was Head of the Benthic Ecology Laboratory. His research interest is focused on the functional ecology of benthic communities inhabiting shallow and deep waters of the Gulf of Mexico and the Pacific Ocean. His scientific production includes 137 peer-reviewed articles, 15 book chapters, and two edited books, receiving over 1600 citations. He has led more than 40 ocean surveys supported by national and international research institutions, and has served as a consultant on environmental issues to UNESCO, OEA, Guggenheim, Fulbright, Chevron-Texaco, Smithsonian, Philadelphia Academy of Science, and the Natural Environmental Research Council, UK. New genera and species of marine invertebrates have been named after him to honor his scientific career. His disciples include three postdoc, 12 PhD, 14 MSc, and 12 BS degrees He is a regular member of the Mexican Academy of Science, Sigma-Xi Society, and holds the highest ranking in the National Research System in Mexico.

# Contents

Mitochondrial Group I Introns in Hexacorals Are Regulatory
Genetic Elements
*by Steinar Daae Johansen and Åse Emblem*

# Preface

The benthos zone constitutes that part of the marine ecosystem that has the maximum biodiversity and biomass of planet Earth. The benthic researcher's state of the art has received a tremendous impulse thanks to the use of cutting-edge technology such as remotely operated vehicles, autonomous submersibles, and sophisticated molecular techniques that have progressively unraveled the secrets of the marine benthic realm. We are now entering a new and more ambitious research era studying the processes that make possible the existence of life from the coastal zone down to the most inhospitable deep habitats (hydrothermal systems). The Intergovernmental Panel on Climate Change has strongly emphasized the need to assess the risks and vulnerabilities of coastal and deep-water systems. Predicted global atmospheric warming causes changes in the circulation pattern of major ocean currents, its surface thermal regime, and the average sea level. Our current knowledge of the adaptations of organisms exposed to the above disturbances remains limited as does the resilience capacity of the benthic system to global-scale change phenomena.

The chapters included in this book represent clear evidence of the interdisciplinary efforts focused on the assessment of the effects of global-scale environmental changes upon benthic communities. To accomplish the general objectives of our book, its content has been divided into two sections: I. Marine and Limnetic Ecosystems and II. Coral Reef Ecosystems.

*Marine and Limnetic Ecosystems*

The introductory chapter is followed by the first chapter in this section, which contains an excellent review of a long-term study of the benthic communities distributed along the southwestern Brazilian continental shelf. The author presents the reader with a thorough description of the hydrographic, geochemical, and biological processes governing the structure and functioning of complex benthic assemblages in the study area. The analysis is mainly focused on the infaunal polychaeta's component and other epifaunal species. The text is rich in valuable information dealing with the ecological response of a benthic system exposed to seasonal hydrographic processes, and how the dominant infaunal–epifaunal elements adapt themselves to take advantage of the quantity and quality of organic matter supplied by oceanic and coastal sources.

This chapter is the product of a multidisciplinary research effort promoted by one of the most prestigious oceanographic institutions in Latin America. The outcome of such effort has been a series of outstanding scientific contributions dealing with several aspects of functional marine ecology that have significantly contributed to open new research fields like ecological chemistry and the health of marine ecosystems.

The second contribution in this section offers the reader an in-depth review of the vital importance of using stable isotopes in biogeochemical studies focused on elucidating the source and flow of carbon and nitrogen in contrasting marine environments of the south and eastern Gulf of Mexico. To achieve this goal, the author has appropriately chosen anthropogenically impacted areas and other sites that presumably can be considered undisturbed (pristine). Such regions include hydrocarbon seep sites, coastal zone environments, oil fields, coral reefs, and deep-sea habitats.

The author of this chapter underlines the usefulness of stable isotopes as invaluable chemical tools to study the cycling of natural and anthropogenic gaseous, dissolved, and particulate compounds in the marine ecosystem. The approach of employing different environmental scenarios within the Gulf of Mexico allows the reader to understand the complexity of processes involved in the transformation of organic carbon from primary producer to its eventual incorporation in the benthic trophic web. The isotopic signatures of stable carbon and nitrogen determined in sediments and animal tissues significantly contribute to discerning the nature of the carbon source sustaining a benthic or planktonic community as well as the possible trophic web structure. Terrestrial, oceanic, biogenic, or pyrogenic sources and even paleoclimatic events can be analyzed by a dual carbon and nitrogen stable isotope approach.

Limnetic systems constitute a research subject that has unfortunately received scarce attention in Latin America. Several factors make the study from Peru particularly attractive for the general public: the high unique environmental conditions of the Andes and the deleterious effects of human impacts in pristine habitats.

*Coral Reef Ecosystems*

The section on coral reef ecosystems reveals the current concern of the world community for the preservation of these unique environments. Initially, an informative review on a global environmental issue concerning the deleterious effects caused by the industrial development on climate change and the health of the oceans is included. In this chapter, the authors focus their attention on the acidification of the oceans (AO) produced by the injection of $CO_2$ into the atmosphere induced by anthropogenic activities. To illustrate the AO, the authors have selected specific cases of calcareous hydroids, analyzing the implications of pH fluctuations on essential processes in the construction of calcifying exoskeletons such as biomineralization and skeletogenesis.

The adoption of a biochemical approach contributes to examine the critical role that $Ca^{2+}$ plays in cortical reactions in eggs and muscle contractions, as well as in larval metamorphosis. Calcification rates, oxidation stress, and symbiont response are discussed by the authors who review most of the current literature on these subjects. The authors express a word of caution in extrapolating the outcome of experimental or controlled studies on organisms whose metabolism depends on $Ca^{2+}$ physiology.

This section includes a second contribution in which the authors offer the reader a well-documented overview of the current status of the mesophotic coral habitats in the Mediterranean Sea. Undoubtedly, the Mediterranean represents one of the best-known marine ecosystems in the world. However, as the authors clearly state, there are still some issues that need to be solved. Perhaps one of the most pressing themes is the conservation and protection of fragile subhabitats unexplored due to their inaccessibility with traditional observational strategies. The availability of cutting-edge marine technology (remote sensors) has opened a whole new world of opportunities to assess the invaluable ecological services rendered by deep benthic communities. It is emphasized that the management and conservation of vulnerable marine ecosystems ought to rely heavily on a monitoring program in which political and socioeconomic issues are essential in its implementation.

Another issue of significant importance in the inner ecological balance of a reef community is the role played by invasive species. This subject is dealt with in detail using as an ecological model the sun coral as a multistrategist invader.

Sun corals, *Tubastraea coccinea* and *T. tagusensis*, are invasive scleractinians that rapidly colonize and dominate large areas of rocky substrates, causing alterations in the structure of native benthic communities. Described in French Polynesia and the Galapagos Islands, respectively, both species altered their natural dispersal route as a result of the increase in transport vectors associated with oil exploration, reaching pantropical distribution in a few decades. The impacts that these corals have on marine coastal ecosystem diversity of different regions, affecting tourism, fishing activities, and artisanal shellfish markets, promoted, in recent years, an explosive increase of actions pointed at controlling these bio-invaders. The application of state policies and the creation of international organizations and NGOs explicitly dedicated to the management of areas invaded by these corals are evidence of this. A growing concern on the environmental consequences of sun coral invasions is also reflected in the academic field, with an accelerated production of scientific articles dedicated to disseminating information on ecological aspects responsible for colonization success and its rapid dispersion. Supreme chemical defense, rapid growth, growth strategies to avoid suffocation by other organisms, capacity for regeneration and reorganization of tissues ("bailout" strategy), variety of reproductive strategies, massive production of larvae, selective behavior prior to recruitment, ability to settle on artificial substrates, generalist feeding strategies, and tolerance for variations of environmental factors are some of the aspects that are further commented on in this chapter.

The section ends with a very innovative approach to the study of coral reef ecology. This final chapter unravels the structure, function, and biological role of catalytic RNA introns present in mitochondrial genomes of hexacorals. We learn from this study that the mitochondrial genomes of hexacorals are remarkable for encoding complex catalytic RNA belonging to the Group I family of self-splicing introns. Some introns are obligatory and present in all investigated species and specimens belonging to all five orders of hexacorals: stony corals, mushroom corals, black corals, colonial anemones, and sea anemones. Others appear as mobile genetic elements with a sporadic taxonomic distribution pattern. In this chapter, new exciting findings of hexacoral mitochondrial genomes, with a specific focus on catalytic RNA introns, are reviewed and discussed. How these autocatalytic RNA elements interfere with mitochondrial RNA processing and mitochondrial function is not well known. However, studies indicate new regulatory roles as non-coding RNA that may sense environmental changes.

Additionally, this chapter contains an interesting phylogenetic analysis of hexacorals employing the state of the art in molecular biology. The authors conduct a complete review of the mitochondrial gene organization and expression of five hexacoral orders, focusing mainly on mitochondrial genome (mtDNA) sequencing, and in the presence of self-catalytic Group I introns. The biological significance of mitochondrial introns as purely genetic elements and the possibility that introns could play some regulatory function are also discussed. The authors highlight the likelihood that hexacoral introns may have a fungal origin. In closing, they offer the reader an update on the characterization of mitochondrial genomes of 200 available mtDNA sequences of all five hexacoral orders: Actiniaria, Zoantharia, Scleractinia, Corallimorphoraria, and Antipatharia.

The message transpiring from the contributions contained in this book is that interdisciplinary exercises to assess environmental changes on a global scale should be focused on specific benthic ecosystems: wetlands, coral reef systems, oyster banks, shelves, and deep-sea communities. This action would generate possible scenarios of change on its general composition based on the evaluation of community parameters employing, in this endeavor, the best technology at hand.

As editor of this book, I would like to express my deepest gratitude to the contributing authors for their insightful views on a variety of research topics. A special word of appreciation to the IntechOpen editorial staff for their vital support throughout the gestation of this book.

**Dr. Luis A. Soto**
Instituto de Ciencias del Mar y Limnología,
UNAM,
Ciudad Universitaria,
Mexico

Section 1

# Marine and Limnetic Ecosystems

# Chapter 1

# Introductory Chapter: The Benthic Realm

*Luis A. Soto*

## 1. Introduction

The present book is an unpretentious editing venture to fill the gap in our current knowledge on the ecological implications caused by anthropogenic disturbances upon benthic communities in several regions of the world ranging from the Western Atlantic, the Mediterranean Sea, and the Eastern Pacific Ocean, including the pristine environments of the Andes in South America. The common goal of the contributing authors in this book was to unravel the complex processes that make possible the life existence of bottom-living animals in different environmental scenarios. In order to achieve such a goal, the authors focus their attention on the emerging issues inherent to the global climate change or the pollution of aquatic systems. These are all themes that might be of interest to scientists active in a wide range of oceanographic subdisciplines. Well-established researches would appreciate the innovative approach adopted in each chapter of the book, which extends from the ecosystem level to refine molecular interpretations.

## 2. The benthic realm

Benthic organisms are excellent bioindicators of adverse conditions in marine ecosystems. Their sedentary lifestyle, distribution patterns, and community properties may reveal significant changes in their structure and functioning, caused by natural or anthropogenic disturbances. They can reflect the long-term effects of various sources of pollutants since these remain sequestered in sediments for long periods. Both benthic macrofauna ($>500$ μm) and meiofauna ($42$–$500$ μm) are ideal candidates to establish comparative analyses to study the magnitude of an environmental disturbance between "altered" and "unaltered" sites.

Benthic communities in shallow environments play an essential role in maintaining the ecological balance of tropical coastal systems. They are also closely linked to the socioeconomic development of human populations because their diversity and biomass include biotic resources of commercial and industrial importance. The Intergovernmental Panel on Climate Change (IPCC) has expressed concern about the risk and vulnerability of coastal systems, which may arise from disturbances in the marine environment caused by the increase in atmospheric temperature and sea level. Both factors have been associated with hydrometeorological phenomena (storms and hurricanes), whose consequences have been floods, coastal erosion processes, and the alteration of habitats such as wetlands, reefs, and coastal lagoons. In contrast, the increase in the concentration of $CO_2$ in the atmosphere has been correlated with the acidification levels of the oceans. The balance in the deposition processes of $CaCO_3$ can mean a severe alteration for all the benthic organisms that build their exoskeleton based on $CaCO_3$.

Benthic communities in tropical environments are particularly vulnerable to processes that change the thermohaline regime. Bottom-dwelling organisms are exposed to dilution or salinization, eutrophication processes, as well as to alterations in the deposition of $CaCO_3$. Undoubtedly, one of the phenomena of the most significant concern is coral bleaching as a result of the disruption of the symbiotic relationship between algae and zooxanthellae, attributable to an increase in ambient temperature. In recent years, there has been an exponential increase in the number of publications on the biological effects of ocean acidification (OA), and several recent reviews have covered this topic. The importance of the combined and frequently interactive impacts of multiple stressors (such as temperature, low oxygen, and pollutants) is now recognized, also the potential for multigenerational adaptation. Experimental research confirms that survival, calcification, growth, development, and abundance can all be negatively affected by acidification. However, the scale of response can vary significantly for different life stages among taxonomic groups and according to other environmental conditions, including food availability. Volcanic $CO_2$ vents can provide useful proxies of future OA conditions allowing studies of species responses and ecosystem interactions across $CO_2$ gradients. Studies at suitable vents in the Mediterranean and elsewhere show that benthic marine systems respond in persistent ways to locally increased $CO_2$. At the shelf edge, the ongoing shoaling of carbonate-corrosive waters (with high $CO_2$ and low pH) threatens cold-water corals, in particular *Lophelia pertusa*, in the Northeast Atlantic. These reefs are rich in biodiversity, but we have a poor understanding of their functional ecology and their reactions to the combined effects of future ocean acidification, warming, and other stressors.

Another condition of a critical nature for benthic organisms is the excessive nutrient load discharged by rivers and lagoons into the environment adjacent to the continental shelf. This process is causing the appearance of areas of hypoxia on the seabed, whose epifaunal diversity decreases significantly. Presently, we are more conscious about the severe physical disturbances on the continental shelf, coral reef, or wetland communities that can leave sequels lasting up to more than a decade. Furthermore, at the same time, our inability to predict and prevent disastrous ecological events has become more evident due to our restricted knowledge of biological diversity, stability, and the resilient capacity of benthic environments.

## Author details

Luis A. Soto
Instituto de Ciencias del Mar y Limnología, UNAM, Ciudad Universitaria, Mexico

*Address all correspondence to: lasg1946@gmail.com

IntechOpen

# Integrative Approach to Assess Benthic Ecosystem Functioning on the Southwest Brazilian Continental Shelf

*Ana Maria S. Pires-Vanin*

## Abstract

Continental shelf is a highly dynamic system controlled by water mass interactions, biogeochemical processes, and biological production of organic matter. Climatic and hydrological processes originate large variability in many scales of time and space that are responsible for its typical unsteady status, mainly at shallower depths. The southeastern Brazilian continental shelf is an important economic area that houses the commercial Port of Santos, the Petrobras oil terminal in São Sebastião, and fishery activities. This concise chapter explores the relationships of the benthic community structure facing a complex physical environment allied to human influences. It is built on previous studies developed in the southeast Brazilian continental shelf from the past 25 years. The shelf benthic system is governed by seasonal pulses of primary production promoted by the South Atlantic Central Water bottom intrusion and coastal upwelling allied to the passage of winter cold fronts. Self-structuring benthic community is achieved by the mobility of the organisms, feeding activity, and biogenic transformation of the habitat due to bioturbation.

**Keywords:** water circulation, organic matter flux, community structure, biodiversity, anthropogenic influence, southeast Brazilian continental shelf, southwestern Atlantic

## 1. Introduction

Continental shelf is an extraordinary place for life in the oceans and vital for life support for the planet. Extended periods under high autotrophic biomass and primary production make the area the most productive in the oceans. Despite occupying an area of about 8.9% of the world's ocean, coastal ecosystems generate nearly 25% of the global biological productivity and more than 90% of total fish catch. Seasonal wind-driven water masses promote intense suspension of bottom sediments with consequent rapid return of nutrients to the euphotic zone. Here, the physical transport and biogeochemical transformation processes affect the fluxes of nitrogen and carbon into and out of the system. The relative shallowness of the shelf facilitates the recycling process and is the structural cause of the high biomass found. Indeed, the major part of the atmospheric carbon fixation through photosynthesis occurs in potentially fertile shelves where it becomes incorporated to pelagic and benthic organisms besides bottom detritus.

Continental shelf surrounds every continent and represents the submerse extension of the land. With shallow seas associated forms a dynamic transitional system between the shoreline and deep sea. Width is variable and dependent on local topography with some areas more extensive than others. Mean values are about 70–80 km, and oceans with passive continental margins, like the Atlantic Ocean, present broader shelves than those of active tectonic margins, as the Pacific Ocean. In Brazil, the equatorial northernmost Amazon Shelf is about 330 km wide, whereas in the northeast coast, on parallel 14°S, the narrowest shelf is about 10 km wide [1].

The shelf is a low-sloping platform, with gradients lower than 1:1000 (1 m of decline for 1000 m of extension). However, local variability occurs due to the presence of canyons, valleys, and channels formed mainly during glacial and interglacial periods when sea level fluctuated. The coastline is the landward limit of the shelf that increases in depth to about 100–200 m where the gradient abruptly changes to about 1:40 forming the slope. The shelf break marks the offshore limit of the continental shelf.

Shelves can be divided into different areas according to distance from the coast. Generally, two areas are present, the inner or coastal shelf and outer or external shelf. Sometimes, depending on the shelf's width and hydrological regime, a middle region may appear between the two. The shelf division occurs due to differences in topography, hydrology, or sediment type, and there is no abrupt change between habitats when the frequent species overlap.

Climatic and hydrological processes act intensively on the shelf in several scales of time and space, and consequently, the environment is highly dynamic. Another important characteristic is that stability increases with distance from the coast and depth. Depth is a driving factor, but many others contribute to coastal instability as the seasonal change in temperature and salinity, water mass circulation, waves and storms, type of sediments, rivers inflow with chemical and geochemical alterations, and light. For benthic communities, the type of sediment, food availability, and benthopelagic coupling are essential among other biological and environmental interactions.

Sediments present in shelves are continental in origin and transported mainly by rivers but also by glaciers and winds. Light intensity may extend down 200 m, favoring photosynthesis and plant growth in both the water column and at the sea bottom, with consequent abundance and diversity of benthic life. Also important are the nonliving resources on the seabed such as the oil and gas resources. A great part of the petroleum production nowadays has been drawn out from the shelf.

The loss of marine diversity is higher in shallow coastal areas as a result of conflicting uses of coastal habitats [2]. It is closely connected with ocean pollution and acidification and results from man's interference. More than 50% of the human population lives near the coast, and the intense development of cities and use and abuse of marine waters and bottoms threaten the integrity of shelf systems. Sustainable usage of marine shelf systems continues to be imperative in addition to the living resource management. The pressing need for estimating the species diversity has been a significant asset for conservation programs, and several and useful tools were developed in the last few decades for that. Taxonomic sufficiency [3] and biotic indices among others allow a rapid diversity and structural assessment of the benthic communities of tropical and subtropical areas scarcely studied [4].

## 2. The southeast Brazilian continental shelf: physical environment and nutrient sources

The Brazilian coastline extends for more than 8500 km along the South American continent. It goes from the country's equatorial north to the temperate south, between latitudes 4°N and 34°S, and represents one of the world's longest continental coasts. Narrow in the northeast (c.a. 10 km at 14°S) and wide in the southeast (c.a. 180 km), coastal shelf presents a variety of ecosystems and habitats that brings expressive biodiversity and endemism to the region. Mangroves, coastal lagoons, and coralline calcareous algal reefs are important ecosystems of the coast, but marine sediments by far provide the largest area for benthic plants and animals. Indeed, after the ocean water column, marine sediments constitute the second biggest habitat on the planet.

The southeast Brazilian continental shelf (SBCS), or south Brazil bight (SBB), is one out of six characteristic physical environments found in the Brazilian continental shelf and the most studied (**Figure 1**). Its coastal limit lies between 23°S and 28.5°S approximately, and the inner, middle, and outer shelves are present on the extensive sea floor and separated by slight declines from each other. Broadly, sediments are distributed in strips along the coastline. Terrigenous bottoms predominate on the inner shelf in contrast with the outer shelf where carbonate sediments are the principals. Inner shelf and proximal bottoms of middle shelf are composed by sand, but near 70 m a sharp change occurs on the middle shelf due to the presence of a large deposit of silt and clay. Based on the bottom topography [5] and benthic macrofauna distribution [6], the north shelf of the southern Brazil bight was divided into two major areas, inner and outer shelf, separated by the 50 m isobath due to the strong coupling of macrofauna and sedimentary variables. Such division is valid for benthic animals with restricted locomotion and lifestyle dependent on geochemical characteristics of sediment grains.

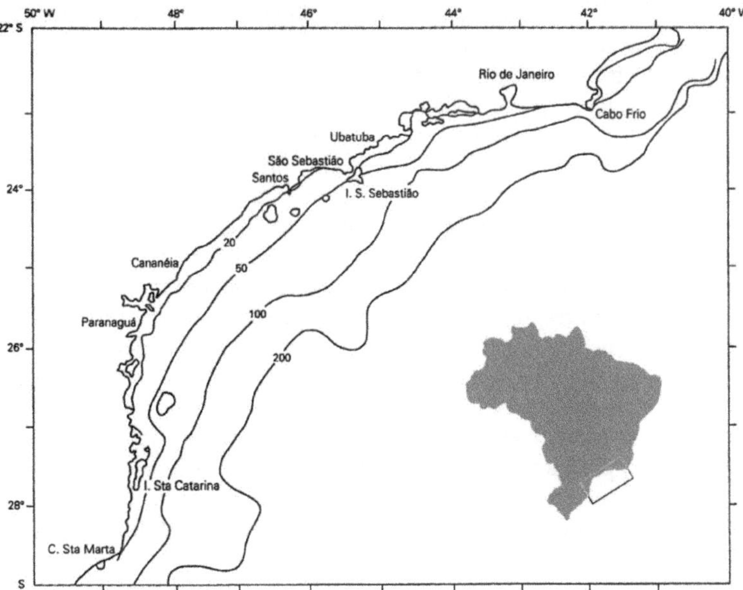


**Figure 1.**
*Location of the southeast Brazilian continental shelf. The isobaths depict the brad shelf configuration and the geographic position of the studied sites.*

In the major part of the southeast Brazilian continental shelf, water movement is driven in different time scales by wind, the Brazil Current (BC), and tides [1]. The Brazil Current is part of the southward branch of the South Equatorial Current when it approaches the coast of South America between 7 and 17°S [7]. It flows southwestward along the shelf break to the Subtropical Convergence, between 33 and 38°S. In wider shelves the Brazil Current approximates to the coastline and fills at least outer shelf [1].

The Brazil Current transports three water masses with characteristic thermohaline properties that interact along the shelf and shelf break according to the large-scale wind field: Tropical Water (TW), relatively warm (T > 20°C) and salty (S > 36); South Atlantic Central Water (SACW), relatively cold (T < 20°C) and low saline (S < 36); and Coastal Water (CW), warm and low saline (T > 20°C and S < 36) [8].

Tropical Water occupies the surface of the Tropical South Atlantic and is known as oceanic or offshore water. South Atlantic Central Water is oceanic in origin and formed by subduction of surface waters of the Subtropical Convergence. After a complex flow, it reaches the Brazilian coast most probably at Cabo de São Tomé (22°S) from which it is transported southwestward by the Brazil Current. It flows along the slope and can reach the shelf to compensate the Ekman transport of surface waters offshore caused by northeast winds. Winds are intense in the austral summer when South Atlantic Central Water intrudes from slope to shelf shallower depths in a cross-shelf transport. Continental waters (from rivers, estuarine plumes) mix with South Atlantic Central Water and Tropical Water resulting in the Coastal Water predominant on the inner shelf [1].

The Brazil Current presents also meandering and formation of mesoscale eddies (nearly around 100 km in diameter) in its frontal edge that facilitates the ascension of nutrients from deep areas and causes upwelling at the shelf break, with the consequent fertilization of large areas of outer shelf. When this process occurs, the regenerated production characteristic of oligotrophic open seawaters is temporarily substituted by new production based on input of new nutrients to the area. Besides the shelf break resurgences, coastal upwellings occur near to the coast and have local effects only. It is caused by northeast winds that when strong and intense deviate the surficial waters to offshore with the consequent ascension of water rich in nutrients from the South Atlantic Central Water. The most studied coastal resurgence in the southeast Brazil bight is that of Cabo Frio shelf (23°S), north of Rio de Janeiro State. Here shelf is narrow (nearly 90 km wide) and steep, which facilitates local water resurgence. As wind-driven the shelf upwelling of SACW is more recurrent and constant in the period from austral late spring to early autumn.

Water masses dynamics linked to the presence of vortices of local and mesoscales are in great part responsible for water column fertilization with direct impact on planktonic and benthic shelf communities. Waters acting on southeast Brazil bight are a result of the three water masses mixed in variable volumes. Coastal Water plus the seasonal intrusion of South Atlantic Central Water coastal wards enhance abundance and diversity of biological communities on the shelf as they feed from the new production at the base of the euphotic layer, a labile and fresh organic matter colonized by heterotroph bacteria. SACW is rich in minerals from deep water and when breaks the thermocline during its ascending to surface, reaches the photosynthetic layer and fertilizes shelf from bottom to mid-waters. The biological consequence is an expressive primary production and improvement of the local food web.

Continental fertilization of seawater comes from medium- or small-sized outflows that contribute to the low saline waters of the inner shelf. Large rivers or estuaries are absent in the southeast Brazil coast (SBB). The coastal lagoon system of

Cananeia in south SBB and Santos Estuary in the central part are important freshwater local inputs. Coastal currents are parallel to the coast, mainly northeast directed, and can be intensified in speed by winds of cold fronts more frequent and stronger during austral winter. These fronts can resuspend bottom sediments and bring the particulate organic matter to the water column promoting the recycling of nutrients and enhancement of the benthopelagic coupling. Wind is the main forcing agent on water circulation in SBB, while tidal currents have a negligible influence [1].

According to the characteristics and dynamics of the water masses present, hydrographic fronts may occur on inner and outer shelves. The front is the water masses interface with distinct physical, chemical, and biological characteristics. The presence of SACW in contact with Coastal Water and Tropical Water in the euphotic zone configures a frontal zone. Also, fronts are horizontal gradients of temperature and salinity formed due to differences in depth, wind direction and intensity, and water density, among others. Detection of thermal fronts, for instance, can help to identify zones of ecological importance for marine fauna and to better understand habitat dynamics as a function of its spatial and temporal extent and variability [9]. Evidence of the influence of thermodynamic fronts on benthic megafauna living in central and northern southeast Brazil bight will be presented later in this chapter.

On the southeast Brazil bight, the inner side of the Deep Thermal Front tidal circulation maintains a mixed layer from surface to bottom in contrast with the side outward from the front that is constantly stratified. Especially in summer, when the offshore SACW intrudes coastward, the physical stratification is enhanced though the shelf. As a consequence of the two-layered water column establishment, the changing of substances and organisms between the surface and the seafloor is inhibited. The 20°C isotherm indicates the limit beyond that South Atlantic Central Water dominates the shelf bottom layer. Similar to temperature, salinity has the Shelf Hyaline Front (SHF) originated between the coastal low saline-mixed waters and the stratified high saline waters from South Atlantic Coastal Water and Tropical Water on the outermost shelf.

The shelf eutrophication promoted by the upwelling of deep water is intermittent and more frequent in summer. During winter SACW retreats to the shelf break more often due to the change in the direction of prevailing winds. In this case, the shelf's bottom is filled with the warm less enriched Coastal Water, while the oligotrophic Tropical Water dominates at the surface.

To summarize, the southern Brazil bight has oligotrophic upper waters (Coastal Water and Tropical Water) in most parts of the shelf in the absence of an external source of nutrients. When the environment is perturbed by SACW intrusion, rich in nutrient salts, an increase in phytoplankton biomass occurs due the presence of new species better adapted to compete in the new condition. On the inner shelf, phytoplankton biomass data are in the range of coastal oligo-mesotrophic areas, and values between 0.16 and 6.42 mg Chl-a m$^{-3}$ were observed for São Sebastião shelf and similar neighboring places [20]. The presence of cross-shelf intrusions, meanderings, and resurgences of SACW permits entrance of new nutrients at the base of euphotic layers of both inner and middle shelves in summer and outer shelf in the winter. The chlorophyll maximum layer is formed in subsurface following the SACW superior limit. The continental shelf is then fertilized in summer by large autotrophic plankton, mainly diatoms, and local primary production frequently enhances several times. In the stratified waters of Ubatuba shelf, the maximum value of chlorophyll-a equal to 14.7 mg m$^{-3}$ was found at 18 m depth in the SACW, which is 13 times higher than that obtained at the surface in the Coastal Water [10]. Another example is that of the Vitoria Eddy, Abrolhos Bank, where the increase of nutrients from deep water turns the area nearly 40% more productive than that out of the vortex [11]. The deep chlorophyll maximum (DCM) layer may reach several

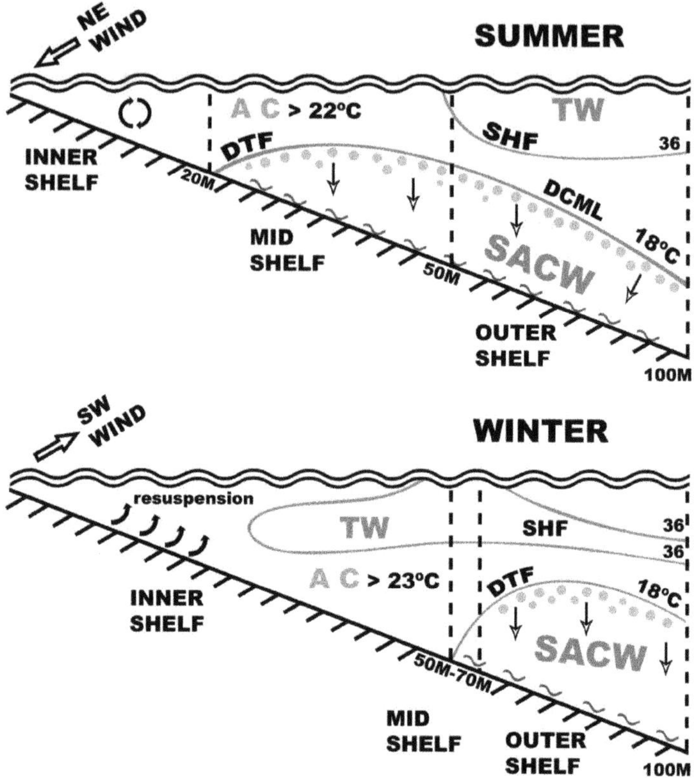



**Figure 2.**
*Diagrammatic model of the main physical and biological processes in the Southeast Brazilian continental shelf in summer and wintertime. Shelf division is based on seasonal hydrodynamics. AC = coastal water, SACW = South Atlantic central water, TW = tropical water, DTF = deep thermal front, SHF = surficial hyaline front, DCML = deep chlorophyll maximum layer, o = phytoplankton cells, ~ = detritus, ↓ = phytoplankton sinking and benthos enrichment, ↑ = resuspension of bottom sediments.*

meters in thickness depending on wind force and shelf depth. The eutrophication benefits from wind strength for reaching shallower depths on the shelf. Also, in summer, more than one event can occur independently on the middle and outer shelves as was demonstrated offshore of Santa Catarina State, southern southeast Brazil bight [12]. New nutrients significantly improved carbon net biomass and exportation of organic matter to benthic system with a consequent increment of secondary production. A diagrammatic model of the biological and physical interactions for the southeast Brazilian continental shelf is presented in **Figure 2**.

## 3. Food supply to the benthic system

With the main mechanisms of shelf eutrophication understood, it was possible to estimate the quantity of organic matter on southeast Brazil bight fuelled to the sediments. Knowledge about the relationships between macrofauna and organic matter input is crucial for understanding the structure and dynamics of benthic communities. The role of the remote source of nutrients represented by the South Atlantic Central Water shelf intrusion has been studied intensively on southeastern Brazilian continental shelf in a multidisciplinary approach. In the São Sebastião Channel (SSC), NE São Paulo State, a clear relationship between high quantities of fresh organic matter and SACW intrusion was observed on bottom sediments [13] and in the Cabo Frio resurgence as well [14].

The quality of sedimenting particles, however, is difficult to be evaluated due mainly to the complexity of intrinsic variables involved and the inexistence of a universal marker for quality. Prevailing oceanographic condition, depth, time and duration, concentration, and heterogeneity of organic content act directly on the organic matter constitution. Considering biomarkers, fatty acids, sterols, and isotopic composition ($\delta^{13}C$ and $\delta^{15}N$) have been frequently used nowadays besides chlorophyll-a and the relation between chlorophyll-a and phaeopigments. Lipid content stocks energy and brings to food high nutritional power and consequently is considered a good indicator of the quality of the particle ingested.

The organic matter concentration and its chemical composition contribute in regulating the metabolism and distribution of organisms as well as the biomass and diversity of communities. Differences in composition show, for instance, the source of the organic matter present on the shelf's bottom. In shallow shelf areas, detritus of continental origin dominates, whereas in middle and outer shelves, organic matter is mainly from oceanic waters.

The impact of food quality on benthic macrofauna communities was evaluated on the São Sebastião Channel (23°30′ to 24°00′ S; 45°05′ to 45°30′ W), São Paulo State, north of southeast Brazil bight [15]. The study searched for differences in species composition, vertical distribution, trophic habits, and bioturbation effects on benthic assemblages (alive bacterial biomass and polychaetes from meio- and macrofauna) submitted to two dissimilar oceanographic conditions, with and without South Atlantic Central Water influence. Different responses for each situation of food input based on fatty acid classes, particulate organic matter quality, and relative contribution of other sources of organic matter to the detritus pool are expected.

However, why do we work with polychaetes and why are there so many ecological studies focused on these animals? The answer is that they are frequently the most abundant infaunal component of macrofauna in sediments, representing 40–50% of the whole macrofauna on coastal and shallow areas of southeast Brazil bight [11]. A wide range of feeding habits and lifestyles give the species capacity to modify bottom deposits by bioturbation, changing geochemical processes such as oxygen and phosphate fluxes [16]. An important part of the benthic research developed on the southeastern Brazilian shelf has been accomplished employing polychaetes as a proxy of the total macrofauna.

São Sebastião Channel is a peculiar area in the southeast Brazil bight inner shelf due to its geomorphology and hydrodynamic complexity. With nearly 25 m of length, it separates the continent from the large São Sebastião Island (SSI). The SSC with a width of 6–7 km and a depth of 20–25 m at the south and north entrances, respectively, narrows to about 2 km in the middle length where it is as deep as 45 m and curves northwest. It functions as a tunnel for winds from the open sea magnifying its strength. In the channel Coastal Water flows from the northeasternmost part of the year. Intense and strong winds in late spring and summer months promote SACW inflow through the channel's south entrance where a paleo-valley runs out on the island side. At this time a well-defined thermocline establishes in the water column with the two water masses running in the opposite direction, the warm low saline Coastal Water at the surface in SW direction and the cold saline South Atlantic Central Water on the bottom in a NE direction. The hydrology is more complex due to a counterclockwise vortex promoting the flow attenuation in the north insular side [17].

The main transport of sediments on São Sebastião Channel occurs from southwest to northeast with a tendency for more intense deposition of silt and clay along

the continental margin and middle part, places of low current speed. The existence
of distinct sediment patches is one of the leading causes associated with the high
benthic diversity found in the area [18]. Another critical factor to be considered
is the chronic oil and sewage contamination present in the central narrower part
of the channel due to the presence of the São Sebastião Harbor, the DTCS large oil
terminal, and the Araçá sewage pipe responsible for discharges of a quarter of the
urban sewage of São Sebastião city. Low current speed makes difficult the dispersion of contaminants that are deposited in the fine sediments below. The resulting
effect is a change in the quality of the bottom environment. Analyses of total
organic carbon (TOC) of sediments in the central area of SSC showed high values
that are indicative of organic enrichment [19]. This condition associated with the
sewage discharge and petroleum-derived hydrocarbon creates a eutrophic environment that puts the benthic species at risk of damage [4, 14]. Indeed, loss of abundance and diversity of species of the whole macrofauna were observed earlier in
that area characterized by an unbalanced community [18]. So, although the waters
of the São Sebastião Channel were described as meso-oligotrophic [20], the bottom
can be considered eutrophic either by natural or by anthropogenic causes.

In the southeastern continental inner shelf, two mechanisms have been evoked
to support the benthic communities along the year. One is associated with SACW
bottom inflow and seasonal enhancement of the quantity and quality of benthic
organic load. The other is present when Coastal Water is the only water mass
flowing in the area. In shallow depths (<50 m) frequent and intense mixing occurs
in the water column especially in winter months due to the passage of cold fronts.
As the input of nutrients is low and constant, the quantity of organic particles
is not a food stressor for the communities, but quality is. In these areas partially
degraded organic detritus with lower nutritional capacity composes the bottom
organic matter. In springtime 2004 only the relative deeper (15 m) north station on
the São Sebastião Channel was under the South Atlantic Central Water influence,
and the quantity of labile organic material peaked to 206.14 $\mu g\ g^{-1}$, a value four
times higher than those found at the same place in autumn under domination of
Coastal Water [15]. On the non-upwelling scenario of the same shallow area, a rapid
loss of the labile component occurred, and the major part of the organic matter is
partially degraded and accumulated as pointed out by the high values of short-chain
saturated fatty acids found [15]. An important aspect of the mid-water upwelling
of SACW is that its effects on benthos enhancement lasts even after the water mass
returns to offshore. The high quantity of the organic matter settled goes to the bottom subsurface layers due to the reworking of macrofauna. In that manner it stays
available in the sediments for a few months [15, 21].

Organic matter quantity and quality is the primary driver for changes in the
structure of benthic communities. However, besides the organic matter load, it is
necessary to consider the trophic group structure and degree of faunal mobility in
the sediments (or bioturbation) for a better understanding of the process. Benthic
fauna work on the food particles through fractioning and moving them into the
sediments and so making the smaller food parts available to the organisms in a constant action/reaction with the environment. Many studies have been developed in
the area of São Sebastião, Ubatuba, and Santos shelves and north and central areas
of the southeast Brazilian continental shelf, with species of total macrofauna [13,
19], polychaetes [15, 18], amphipod crustaceans [23], bacteria biomass, and meiofauna [21]. The results recognized the organic matter quality and quantity as the
main determinants of the structure of benthic assemblages. The fauna seems to be
not food-limited by the quantity of the organic particles loaded, but by their quality
that can alter species composition, abundance, and diversity. The constant input
and prevalence of local partially degraded organic detritus (refractory material)

in the sediments were shown to be significant and able to maintain the benthic assemblages on shallow coastal areas [15, 21]. On the other hand, places under the South Atlantic Central Water influence showed more abundant and diverse benthic communities that are most probably supported by the high proportion of recently produced planktonic organic matter present in the sediments [15, 18, 22].

The relationships between feeding mode and species mobility are complex, and their study helps to understand the functioning of benthic communities. In the São Sebastião and Santos shelves, several studies were conducted with polychaete and crustacean species to identify their trophic guilds and link them with sediment type and organic matter content [23, 24]. Five trophic groups are reported for the SBB shelf and recently were associated with four bioturbation categories, for a better understanding of the functional structure of polychaete assemblages in the São Sebastião Channel and vicinities [15]. In shallow places with predominance of local input of degraded organic material, as the São Sebastião Channel margins and other coastal areas, the diffusive mixing (rapid redistribution of the organic matter within the sedimentary column) is reported as the main process associated with dominance of the subsurface deposit feeders. The result is the disturbance of the whole sediment column by relatively high bioturbation rates. Species composition of the assemblages can vary along the year, but relevant functional changes were not observed in the system, i.e., different species may occur through time but play the same role. Large quantities of small opportunist species occurred in the São Sebastião Channel continental margin, like *Cossura candida*, together with mobile large Sternaspidae, as *Sternaspis capillata*, both subsurface deposit feeders but with different bioturbation behaviors. *C. candida* is a diffusive mixer, i.e., rapidly redistributes the organic matter within the sedimentary column, and *S. capillata* is a conveyor belt transporter, i.e., moves particles of sediment up to the surface during its subsurface feeding or burrow excavation. The species are characteristic of environments under intermediate stress condition [15], and the input of anthropogenic organic matter locally produced seems to support them.

Benthic assemblages behave differently in the presence of SACW's eutrophication. In such places, the major part of the species belongs to the conveyor belt transport category, that is, individuals that promote rapid movement of recently produced organic matter downward in the sediment. The procedure favors both surface deposit feeders and diffusive mixers equally by combining old and fresh organic matter. So, with the pulses of intense and high bottom eutrophication, a modification of the species composition occurs together with functional changing.


## 4. Benthic studies on the southeast Brazilian continental shelf

Between 1985 and 1988, a multidisciplinary oceanographic project was conducted on the São Paulo State northeastern shelf, by the Oceanographic Institute of the University of São Paulo, to understand the structure and functioning of the continental shelf system from the coast to offshore of Ubatuba, north-south Brazil bight [25]. The study detailed the complex hydrodynamics of the water masses and their role on the large episodic input of new nutrients to the shelf and the consequences on pelagic and benthic communities. It also established the founding knowledge about the functioning of the system based on a seasonal local trophic model. This project was the pioneer in southeast Brazilian shelf by assembling researchers of the many branches of oceanography to understand shelf functioning addressed by its physical, chemical, and biological characteristics. Some other multi- and interdisciplinary projects came along the following 25 years and contributed to improving the knowledge by answering questions opened at every study end.

Without any doubt, the central and north parts of the SBB, in front of São Paulo and Rio de Janeiro States, are the best areas studied. Along the long coastline, few geographical features can modify local sedimentary and hydrographic main processes with consequences on benthic communities' structure and distribution. The first one is the large São Sebastião Island, northeast São Paulo State, separated from the continent by a long narrow channel, the São Sebastião Channel that constitutes the second feature. The third modifier is the sudden east to the northeast inflection of coastline in front of Cabo Frio, northeast Rio de Janeiro State, with consequent expressive shelf narrowing (**Figure 1**). Another critical factor to the change in coastline is that caused by the opening of estuaries: the southern Cananeia/Iguape lagoon system, the larger central Santos-São Vicente estuarine complex, and the northern Bertioga.

São Sebastião Island is an important geomorphologic marker of the coastline as it divides the adjacent shelf in northern and southern sectors. The northern sector is more complex due to the irregular littoral of many small bays and islands associated with the irregular isobaths outline. A clear difference exists between sediments from W to SW and N to NE of the island, with a predominance of fine and very fine sands in the southwest and muddy sediments (silt and clay) in the east and northeast. The SSI functions as a physical barrier to marine currents from S to SW linked to the passage of cold fronts in the winter and is a perennial source of sediments and detritus to the region. São Sebastião Channel is a particular area from the inner shelf and was divided into three sectors (central, south, and north) based on sediment type coupled to depth, channel wall declivity, quantity of suspension matter, and dominant hydrographic processes. Regarding Cabo Frio, the change in the direction of coastline in the area favors the approximation of the Brazilian Current to the continent and, associated with strong northeast winds, promotes the coastal upwelling of deep cold waters that modify local food quantity and quality to benthic assemblages, as explained earlier.

Reports on the importance of sediments for benthic species distribution are numerous in the literature. In the central and north parts of southeast Brazil, shelf studies of the megabenthos have shown that hydrothermal dynamics is the driver factor structuring the communities, whereas for macrofauna the variables associated with sediments are the most important. Megafauna is here defined as large organisms captured by fishnets, like crabs, shrimps, and sea stars, and macrofauna are those invertebrates $\geq$ 0.5mm length from both infauna and epifauna of almost all phyla. Except for Peracarida crustaceans (as isopods, tanaids, and amphipods) that protect their eggs and embryos in the ventral marsupium, most of the other benthic species are benthopelagic with initial free-swimming larval stages and posterior bottom settlement. Sediment is then required in the initial and crucial stage of the species life cycle. Some other organisms, as many shrimps are pelagic, but bottom dependent for feeding. On the other hand, several megabenthic species are large agile animals that need to move long distances for feeding and reproduction and, consequently, are affected by water motion. Recently attention has been paid in studying the role of thermohaline fronts on the habitat dynamics in function of its temporal and spatial extensions. One of these studies was developed in the Ubatuba shelf, SBB shelf, and later expanded to São Sebastião and Santos shelves. The results showed a constant temporal and spatial change of habitat between *Xiphopenaeus kroyeri* and *Portunus spinicarpus*, two coupled species linked to the South Atlantic Central Water deep thermal front.

*X. kroyeri*, also known as sea-bob or "camarão sete barbas," is a penaeid shrimp with a length of 9–10 cm and a long curved rostrum distributed along the southeast Brazil bight coastal area, from Rio de Janeiro to Santa Catarina States. The species lives in shallow warm water (warmer than 20°C) 30 m in depth. The swimming crab *P. spinicarpus* inhabits cold (below 20°C) deeper shelf water (from 50 to 70 m to shelf break) with populations extremely numerous at the 18°C isotherm in the South Atlantic Central Water frontal zone. Both species extend or diminish their spatial range of distribution seasonally according to the constant displacement of SACW [26]. Similar results obtained for other vicinal areas showed that the southeast Brazilian shelf is a dynamic habitat for megafauna species supported by plankton-benthic interactions coupled to physical forces as hydro-thermodynamics, winds, and tidal mixing, among the principals. The seasonal variation in abundance of both species on the three studied localities is presented in **Table 1**.

In Cabo Frio and Ubatuba, a study that lasted over 2 years was developed to compare the megabenthic community structure in relation to different physical processes that occur in those areas, the local upwelling in Cabo Frio and the mesoscale South Atlantic Central Water middle depth intrusion in Ubatuba [27]. Density, biomass, and species richness were evaluated on inner and outer shelves in the austral winter of 2001 and summer and spring of 2002. Substantial spatial and temporal changes occurred in species composition and dominance of key species on both areas and suggested the close linkage between megabenthic communities and water masses seasonal dynamics associated with differences in sediment type.

Considering the inner shelf, diversity in Ubatuba was higher than in Cabo Frio, and both areas presented different species dominance also. In Cabo Frio the sea star *Astropecten brasiliensis* was the most abundant in all the periods sampled, even in the presence of SACW thermal front (Summers 2001 and 2002). *P. spinicarpus* appeared on the inner shelf driven by SACW thermal front, but only predominated on that area in spring 2002 (709 individuals/catch) when bottom temperature reached 13.5°C. However, in wintertime under the warm Coastal Water influence, the number of species increased from 10 to 22, and diversity increased from H = 0.3 to

| Species | *Xiphopenaeus kroyeri* | | *Portunus spinicarpus* | | Water mass | |
|---|---|---|---|---|---|---|
| Places | Inner | Outer | Inner | Outer | Inner | Outer |
| | shelf | shelf | shelf | shelf | shelf | shelf |
| Ubatuba[1] | | | | | | |
| Summer 1985 | 0 | 0 | 3281 | 5152 | SACW | SACW |
| Winter 1986 | 5892 | 0 | 34 | 0 | CW | SACW |
| São Sebastião[2] | | | | | | |
| Summer 1994 | 54 | 0 | 5 | 578 | SACW | SACW |
| Winter 1997 | 385 | 0 | 11 | 407 | CW | SACW |
| Santos[2] | | | | | | |
| Summer 2006 | 47 | 0 | 4 | 1675 | SACW | SACW |
| Winter 2005 | 5413 | 0 | 0 | 1224 | CW | SACW |
| Cabo Frio[3] | | | | | | |
| Summer 2002 | 0 | 0 | 43 | 0 | SACW | SACW |
| Winter 2001 | 0 | 0 | 1 | 1968 | CW | SACW |

[1]*Pires [26].*
[2]*Non-published data.*
[3]*De Léo and Pires-Vanin [27].*
*SACW = South Atlantic Central Water, CW = Coastal Water.*

**Table 1.**
*Seasonal distribution of the abundances of Xiphopenaeus kroyeri and Portunus spinicarpus according to the extension of the South Atlantic Coastal Water intrusion on the shelf.*

1.3, with dominance of the crab *Leucippa pentagona* and the gastropod *Buccinanops cochlidium* (= *Buccinanops gradatum*) instead of shrimps, as presented in other parts of the southeastern Brazilian inner shelf. Regarding the outer shelves, differences in species composition between both places were also detected, despite their proximity (only 463 km distant) and permanent SACW influence. In this case, the contrast in sediment type explains the faunistic changes: Ubatuba has a sandy bottom (coarse and medium sands) at 100 m, whereas in Cabo Frio the sediment is silted at the same depth. The hydrodynamic characteristics associated with sediments are responsible for the major part of the shift on the structure of the communities of both areas. This is especially true for the slow-moving megabenthic species as the sea stars and anomuran crustaceans (*Munida irrasa*). As they feed mainly on local macrofauna living on sediments, the quantity and quality of the prey available for feeding depends on the sediment characteristics, as grain size and organic content [26].

Spatial distribution of the communities of benthic macrofauna has been usually related to seafloor characteristics, like topography, sedimentary texture, oxygen content, organic matter, and depth. Studies developed in the south Brazil bight have shown that besides those variables, the oceanographic and meteorological processes (as SACW intrusion and cold fronts, respectively) play an important role also. Spatial and temporal changes in the communities of macrofauna were intensely studied on the São Paulo State shelf, the central part of the SBB. Based on bottom topography, water circulation, sediment deposition, and sedimentary organic matter content, the area was characterized by three subareas: the northern Ubatuba and São Sebastião shelves [13, 25, 26], the central Santos shelf [24, 29] and the southern Cananeia/Iguape shelf. Since sediment was identified to be the structural driving factor for the macrobenthic communities, a detailed explanation of its distribution in the complex continental shelf of São Paulo becomes necessary. The presence near the coast of the large São Sebastião Island, associated with water fluxes from Santos, Peruíbe, and Bertioga estuaries, adds to the sedimentary system complexity.

The regional distribution of the superficial sediments indicated the presence of three regions. In the south region, corresponding to the continental shelf in front of the Peruíbe river mouth, the sediment presents an average diameter corresponding to fine sand with isolated silt patches. In the central portion of the area, the continental shelf of Santos has very fine and fine sands, which form the homogeneous bottom, with muddy sediments deposited in the deeper portions. In the north region, situated north of the São Sebastião Island, the deposition pattern is much more complex; in areas shallower than 60 m very fine sand interspersed with bands of fine sand predominate, while near to the coast, spots of medium and coarse sand overlay the bottom. The major part of the smaller grains is retained into the bays, but some quantity can be carried on the water surface layer by the Coastal Water during the South Atlantic Central Water bottom intrusion and deposited around the 50–60 m isobath at the middle shelf. Another shallow depositional environment is that at E/NE of the São Sebastião Island. The island functions as a shield to waves from the highly dynamic southern frontal systems by changing current direction and diminishing its velocity. Consequently, the finer sediments are deposited behind the island on an area known as "island shadow zone." Considering the outer shelf of the three regions, there is a clear relation between increase of depth and decrease in the sediment granulometry; the 120 m isobath practically delimits the zone of the predominance of sand from that of muddy sediments [30].

Macrofauna of São Paulo continental shelf is numerous, diversified, and firstly distributed according to sedimentary characteristics. Polychaetes show up as the most numerous group collected accounting for 51% of total fauna on São Sebastião shelf [28]. Polychaetes and crustaceans Peracarida were studied to species level in order to understand benthic community structure and function [24, 31]. Working with several benthic marine groups of invertebrates proved to be important for bringing consistency and generality to the results obtained. Broadly, density was significantly higher in sediments with a sufficient content of labile organic matter. Also, species richness was higher on the coarser sediments of the inner shelf, whereas diversity and evenness increase at sites of intermediate depths (40–50 m) on the middle shelf. In contrast, the outer shelf houses deepwater species that live preferentially in muddy sediments as deep as 70–80 m. As an example, the diversity of amphipods on the northern Ubatuba shelf decreased with the increase of sedimentary silt and clay, whereas abundance of the tube-dwellers follows the contrary [31], which shows the role of the muddy belt from the outer shelf for the shift in community composition along the shelf.

Several ecological models were constructed for the southeast Brazil bight with environmental parameters (grain size and angularity, organic matter quantity and quality, temperature, salinity, and depth, among the principals) and biological indicators (as abundance, biomass, diversity, evenness, feeding groups, behavioral groups, microbial biomass) to interpret the benthic species distribution on the shelf. Results demonstrated that the species are grouped into three main areas parallel to the coastline, forming communities with particular physical, sedimentary, and geochemical characteristics and controlled by different species. Three main groups of species characterized three benthic areas, the inner, the middle, and the outer shelf groups that delimit the coastal zone, the intermediary zone, and the deep zone. The most striking differences occur between the inner and outer shelf groups; the middle shelf functions as an ecotone with species from both areas.

The coastal zone (12 to 30–40 m) includes shallow warmwater species related to well-sorted and very fine sand bottoms with labile and refractory organic matter mixed, subjected to a strong hydrodynamics associated with water masses and cold fronts; density and diversity are generally high, and the fauna is composed mainly by r-strategists as *Prionospio dayi* (polychaete) and *Ampelisca paria* (amphipod). The deep zone (>70 m) sustains cold-water species living in poorly sorted silt and clay bottoms with high organic matter content; density is variable but frequently high. The robust correlation of density and evenness with depth indicates that the deep zone is a more stable environment than the coastal zone and frequently supports an abundant fauna that can reach densities of 958 individuals per m$^{-2}$ [28]. Some discriminant species here are the large carnivores *Sigambra* sp. and *Aglaophamus* sp., the subsurface deposit feeders *Petersenaspis capillata* and *Leitoscoloplos kerguelensis* (polychaetes) [29], and the burrowers *Pseudoharpina dentata*, *Urothoe* sp., and *Heterophoxus videns* (amphipods) [24]. On the other hand, as a transitional region, the intermediary zone is usually characterized by high values of species richness, diversity, and evenness besides chlorophyll-a content correlated with density, which indicates fresh organic matter input. These findings evidence the continuous organic load and enrichment of sediments in the area and no food limitation for the fauna. This is an unstable environment with annual ranges of temperature between 17.2 (summer) and 21°C (winter) as observed in Santos' shelf.

The coastal zone may present patches of finer sediments in front of river outflow or of physical coastal modifiers, as in the case of San Sebastião Island that creates a "shadow zone" behind. The accumulation of such very different sediments, muddy and rich in refractory organic matter, modifies the inner shelf bottom and creates a local environment with a new sedimentary process and particular geochemical properties. As a consequence, an abrupt change occurs and disrupts the sandy community pattern present in the rest of the area. This is the case of the shallow bottom in front of Peruíbe river mouth, south of Santos shelf. The community of the less saline muddy sediment (salinity of 34.6 to 35 in winter (2006) and 33.1 to 33.9 in summer (2007)) is characterized by *Mediomastus* sp. and *Magelona posterelongata;* the first one is a subsurface deposit feeder and a conveyor belt transporter, and the second is an interface-feeder and surface depositor. Species of *Mediomastus* are always very abundant in systems constantly exposed to high organic load [29]. Indeed, the terrigenous input of detritus to coastal waters in São Paulo shelf is permanent and increases in summer, the period of more intense rains and with a fresh food supply associated with the South Atlantic Central Water intrusion. As an interface-feeder, *M. posterelongata* is stimulated just after the deposition of freshly organic matter [32], but is exceeded by the *Mediomastus* sp. that feed on the abundant and long-lasting degraded material [15, 29]. On the other hand, species of polychaetes of the sandy group were mostly tubicolous and surface-feeders and associated with the high content of chlorophyll-a and cold water in summer, which suggests pulses of fresh organic matter to those communities. The structure of the three shelf species groups just discriminated seems to be resilient and time-stable. Several ecological indices used, as well as the different size strata of benthic animals (macro- and megafauna) and different taxonomic groups used, indicated the existence of the same organization independent of species composition or season. When a shift of species occurs, as in the seasonal changes, the new species will play a similar functional role of the substituted, as observed for polychaetes in São Sebastião shelf.

## 5. Conclusion

The southern Brazilian continental shelf ecosystem is characterized by both high species diversity and complex biotic interactions among the component species. The region is physically controlled by the dynamics of three water masses. One of them, the South Atlantic Central Water intrudes from the shelf break to the coast seasonally bringing nutrients to the euphotic zone and, consequently, enhancing primary productivity. The thermal front formed in the frontal zone between South Atlantic Central Water and the shallow Coastal Water is responsible for the concentration of an extremely dense population of the swimming crab *P. spinicarpus* that moves according to the SACW dislodgment on the shelf. Density of macrofauna in south Brazilian bight is moderate and linked to seasonal pulses of labile organic matter in the middle and inner shelves and to water mixing processes that resuspend bottom sediments with old and fresh detritus for recycling. This suggests that the quality besides quantity of organic matter available as food in sediments is of great importance for structuring the macrobenthic communities. Biomass was usually low when compared to other similar environments elsewhere and probably is related to the characteristic mesotrophy of the shelf waters. Changes in macrofauna density and biomass seems to be independent of the periods of high food availability, but the differential quality of the sediments can change community species composition by differences in trophic habits and mobility behavior. Diversity is high, mainly on the middle shelf and outer shelf. Dominance of few species is a characteristic of the inner and outer shelf zones. The reciprocal interaction between sediments and species helps in maintaining the community dynamics through time.

## Acknowledgements

I would like to thank Ricardo Pires Vanin for the graphic design of the figures; to colleagues, students, and all the people who collaborated somehow for data achievement; and to Fundação de Apoio à Pesquisa do Estado de São Paulo (FAPESP) and Conselho Nacional de Pesquisa e Tecnlogia (CNPQ) for giving me financial support for several research projects whose data were in part presented here.

## Author details

Ana Maria S. Pires-Vanin
Instituto Oceanográfico, University of São Paulo, São Paulo, Brazil

*Address all correspondence to: amspires@usp.br

IntechOpen

## References

[1] Castro BM, Miranda LB. Physical oceanography of the Western Atlantic continental shelf located between 4°N and 34°S. In: Robinson AR, Brink KH, editors. The Sea. New York: Wiley & Sons; 1998. pp. 209-251

[2] Gray SJ. Marine biodiversity: Patterns, threats and conservation needs. Biodiversity and Conservation. 1997;**6**:153-175

[3] Oliver L, Beattie AJ. A possible method for the rapid assessment of biodiversity. Conservation Biology. 1993;**3**:562-568. DOI: 10.1046/j.1523-1739.1993.07030562.x

[4] Muniz P, Venturini N, Pires-Vanin AMS, Tommasi LR, Borja A. Testing the applicability of a marine biotic index (AMBI) to assessing the ecological quality of soft bottom benthic communities in the South America Atlantic region. Marine Pollution Bulletin. 2005;**50**:624-637. DOI: 10.1016/j.marpolbul.2005.01.006

[5] Kowsmann RO, Costa MPA. In: Petrobras, editor. Sedimentação Quaternária da Margen Continental Brasileira e das Águas Oceânicas Adjacentes. Rio de Janeiro: Projeto Remac; 1979. pp. 1-55

[6] Pires-Vanin AMS. A macrofauna bêntica da plataforma continental ao largo de Ubatuba, São Paulo, Brasil. Publicação Especial do Instituto Oceanográfico, São Paulo. 1993;**10**:137-158

[7] Stramma L. Geostrophic transport of the south equatorial current in Atlantic. Journal of Marine Research. 1991;**49**:281-294

[8] Miranda LB. Forma da correlação T-S de massas de água das regiões costeiras e oceânicas entre Cabo de São Tomé (RJ) e a Ilha de São Sebastião (SP), Brasil.

Boletim do Instituto Oceanográfico. 1985;**33**(2):105-119. DOI: 10.1590/S0373-55241985000200002

[9] Millera PI, Xua W, Carruthersb M. Seasonal shelf-sea front mapping using satellite ocean colour and temperature to support development of a marine protected area network. Deep Sea Research Part II: Topical Studies in Oceanography. 2015;**119**:3-19. DOI: 10.1016/j.dsr2.2014.05.013

[10] Aidar E, Gaeta S, Gianesella-Galvão SMF, Kutner MBB, Teixeira C. Ecossistema costeiro subtropical: Nutrients dissolvidos, fitoplâncton e clorofila-a e suas relações com as condições oceanográficas na região de Ubatuba, SP. Publicação Especial do Instituto Oceanográfico. 1993;**10**:9-43. Avaliable from: http://www.io.usp.br/images/publicacoes/n10a03.pdf

[11] Gaeta S, Lorenzzetti JA, Miranda LB, Susini-Ribeiro SMM, Pompeu M, Araujo CES. The Victoria Eddy and its relation to phytoplankton biomass and primary productivity during the austral fall of 1995. Archives of Fisheries and Marine Research. 1999;**47**(2-3):253-270

[12] Brandini FP, Nogueira M Jr, Simião M, Codina JCU, Noernberg MA. Deep chlorophyll maximum and plankton community response to oceanic bottom intrusions on the continental shelf in the south Brazilian bight. Continental Shelf Research. 2014;**89**:61-75. DOI: 10.1016/j.csr.2013.08.002

[13] Arasaki E, Muniz P, Pires-Vanin AMS. A functional analysis of the benthic macrofauna of the São Sebastião Channel (southeastern Brazil). Marine Ecology. 2004;**25**(4):249-263

[14] Carreira RS, Canuel EA, Macko SA, Lopes MB, Luz LG, Jasmin LN. On the accumulation of organic matter on

the southeastern Brazilian continental shelf: A case study based on a sediment core from the shelf off Rio de Janeiro. Brazilian Journal of Oceanography. 2012;**60**(1):75-87. DOI: 10.1590/S1679-87592012000100008

[15] Venturini N, Pires-Vanin AMS, Salhi M, Bessonart M, Muniz P. Polychaete response to fresh food supply at organically enriched coastal sites: Repercussion on bioturbation potential and trophic structure. Journal of Marine Systems. 2011;**88**(4):526-541. DOI: 10.1016/j.jmarsys.2011.07.002

[16] Waldbusser GG, Marinelli RL, Whitlatch RB, Visscher PT. The effects of infaunal biodiversity on biogeochemistry of coastal marine sediments. Limnology and Oceanography. 2004;**49**:1482-1492. DOI: 10.4319/lo.2004.49.5.1482

[17] Furtado VV, Bonetti Filho J, Rodrigues M, Barcellos RL. Aspectos da sedimentação no Canal de São Sebastião. Relatório Técnico do Instiuto Oceanográfico. 1998;**43**:15-31

[18] Pires-Vanin AMS, Arasaki E, Muniz P. Spatial pattern of benthic macrofauna in a sub-tropical shelf, São Sebastião Channel, southeastern Brazil. Latin American Journal of Aquatic Research. 2013;**41**(1):42-56. DOI: 103856/vol41-issue1-fulltext-3

[19] Silva DAM, Bicego MC. Polycyclic aromatic hydrocarbons and petroleum biomarkers in São Sebastião Channel, Brazil: Assessment of petroleum contamination. Marine Environmental Research. 2010;**69**:277-286. DOI: 10.1016/j.marenvres.2009.11.007

[20] Gianesella SMF, Kutner MBB, Saldanha-Corrêa FMP, Pompeu M. Assessment of plankton community and environmental conditions in São Sebastião Channel prior to the construction of a produced water outfall. Revista Brasileira de Oceanografia. 1999;**47**:29-46. DOI: 10.1590/S1413-77391999000100003

[21] Sumida PYG, Yoshinaga MY, Ciotti AM, Gaeta SA. Benthic response to upwelling events off the SE Brazilian coast. Marine Ecology Progress Series. 2005;**20**:35-42. DOI: 10.3354/meps291035

[22] Quintana CO, Yoshinaga MY, Sumida PY. Benthic responses to organic matter variation in a subtropical coastal area off SE Brazil. Marine Ecology. 2010;**31**(3):457-472. DOI: 10.1111/j.1439-0485.2010.00362.x

[23] Muniz P, Pires AMS. Trophic structure of polychaetes in the São Sebastião Channel (southeastern Brazil). Marine Biology. 1999;**134**:517-528

[24] Rodrigues CW, Pires-Vanin AMS. Spatio-temporal and functional structure of the amphipod communities off Santos, southwestern Atlantic. Brazilian Journal of Oceanography. 2012;**60**(3):421-439. DOI: 10.1590/S1679-87592012000300013

[25] Pires-Vanin AMS, editor. Estrutura e Função do Ecossistema de Plataforma Continental do Atlântico Sul Brasileiro. São Paulo: Publicação Especial do Instituto Oceanográfico 10; 1993. pp. 1-245

[26] Pires AMS. Structure and dynamics of benthic megafauna on the continental shelf offshore of Ubatuba, southeastern Brazil. Marine Ecology Progress Series. 1992;**86**:63-76

[27] De Léo FC, Pires-Vanin AMS. Benthic megafauna communities under the influence of the South Atlantic central water intrusion onto the Brazilian SE shelf: A comparison between an upwelling and a non-upwelling ecosystem. Journal of Marine Systems. 2006;**60**:268-284. DOI: 10.1016/j.jmarsys.2006.02.002

[28] Pires-Vanin AMS. Megafauna e macrofauna. In: Pires-Vanin AMS, editor. Oceanografia de um Ecossistema Subtropical: Plataforma de São Sebastião, SP. São Paulo: Editora da Universidade de São Paulo; 2008. pp. 311-349

[29] Shimabukuro M, Bromberg S, Pires-Vanin AMS. Polychaete distribution on the southwestern Atlantic continental shelf. Marine Biology Research. 2016;**12**(3):239-254. DOI: 10.1080/17451000.2015.1131299

[30] Conti LA, Furtado VV. Geomorfologia da plataforma continental do Estado de São Paulo. Revista Brasileira de Geociencias. 2006;**36**(2):305-312

[31] Santos KC, Pires-Vanin AMS. Ecology and distribution of Peracarida (Crustacea) in the continental shelf of São Sebastião (SP), with emphasis on the amphipod community. Nauplius. 2000;**8**(1):35-53

[32] Pearson TH. Functional group ecology in soft-sediment marine benthos: The role of bioturbation. Oceanography and Marine Biology Annual Review. 2001;**39**:233-267

# Stable Carbon and Nitrogen Isotopes in Hydrocarbon and Nitrogenous Nutrient Assessment of S and E Gulf of Mexico Marine Environments: Four Isotope Stories

*Diego López-Veneroni*

## Abstract

Stable carbon and nitrogen isotopes were sampled in representative environments of southern and eastern Gulf of Mexico to trace carbon and nitrogen sources and processes affecting them. Sampled sites include a hydrocarbon seep area, a coastal zone influenced by terrestrial discharge, a productive oil field, a coral reef, and a deepwater environment. In Cantarell oil field, $\delta^{13}C$ and $\delta^{15}N$ values of suspended particulate matters, sediments, and benthic organisms show that the principal carbon source to the benthic food web is the downward flux of upper-layer primary production. In the coastal zone, the isotopic terrestrial signature of suspended particles across the low salinity plume indicates that the terrestrial contribution in nearshore waters is progressively diluted by marine organic matter. Hydrocarbon concentrations and $\delta^{13}C$ values from a Bay of Campeche hydrocarbon seep sediment core suggest that the seep contributes to about 72.4% petrogenic carbon to its surface sediment layer. The $\delta^{13}C$ values in corals suggest a carbon source from fixation by zooxanthellae. In the eastern Gulf, organic carbon ($C_{org}$) and total nitrogen (TN) concentrations and isotopes are indicative of low terrestrial contribution, and the principal long-term nitrogen source to primary producers appears to be nitrate diffusing from the thermocline into the photic zone.

**Keywords:** carbon-13, nitrogen-15, Bay of Campeche, Gulf of Mexico, oil seep, suspended particles, sediments, coral reef

## 1. Introduction

The inclusion of stable isotope measurements in environmental studies has proven useful to discern the source and trace the flow and cycling of natural and anthropogenic gaseous, dissolved, and particulate compounds. Biogeochemically relevant stable isotope pairs, such as $^2H/^1H$, $^{13}C/^{12}C$, $^{15}N/^{14}N$, $^{18}O/^{16}O$, and $^{34}S/^{32}S$, actively cycle between the biotic and abiotic components of ecosystems. These elements are incorporated by organisms into a variety of compounds which are then transformed as they are cycled within the organisms, through the trophic chain and back to the abiotic environ. Although the chemical structure of compounds may be transformed in this flow, the isotopic proportion of an element remains constant or tends to vary in a known proportion, thus providing a means for tracking its source and flow. For example, stable carbon and nitrogen isotopes have been used to trace the source of organic matter into food webs and to establish the trophic structure of ecosystems [1–3]. In general, the stable carbon isotope composition of animals approaches that of its diet, with a small fractionation between them (0.8–1.5‰ [1, 3, 4]), while they are generally about 3‰ enriched in $^{15}N$ compared to its diet [2, 4, 5].

The mass difference between the light and heavy isotopes is responsible for small changes in the physical properties of an element but is paramount to trace its origin and different physically, chemically, or biologically mediated processes it undergoes [6–8]. This isotopic fractionation (or discrimination), although small, is measurable, and, in a mixture of isotopes of the same element, the lighter isotope (such as $^{12}C$) is generally favored in the reaction products, leaving the heavier isotope ($^{13}C$) in the reactant.

The heavy-to-light isotope ratio of a sample is denoted as

$$R = {}^{m_X}\!/_{n_X} \tag{1}$$

where $X$ is an element with a heavy isotope ($m$) and a light isotope ($n$).

The stable isotope ratio of a sample ($R_{sa}$) is expressed relative to the ratio of a universal standard ($R_{std}$) by the $\delta$ notation and is expressed in per mil units (‰):

$$\delta = \left({}^{R_{sa}}\!/_{R_{std}} - 1\right) \times 1000 \tag{2}$$

The fractional contribution of two sources (A and B) with different isotopic compositions in a mixture (M) can be estimated from the isotopic composition of each source by isotopic mass balance [8, 9]:

$$\delta_M = f_A \times \delta_A + f_B \times \delta_B \tag{3}$$

where

$$1 = f_A + f_B$$

The southern and eastern Gulf (S and E Gulf) of Mexico (**Figure 1**) offers contrasting environmental scenarios where stable isotopes can be applied as a tool to discern the origin and flow of organic matter. The principal potential carbon sources to the Bay of Campeche continental shelf, located in the S Gulf, include organic matter deposition from upper-layer primary production, input of terrestrially derived organic matter, oil seeping, and offshore oil extraction. In contrast, organic matter inputs of terrestrial origin in the E Gulf are minor.

In this chapter, carbon bend nitrogen sources and flows in these two contrasting Gulf of Mexico regions are traced by means of stable isotopes. Three studies are used here to exemplify the use of stable isotopes in regions with contrasting nutrient and carbon sources. The carbon apportionment is estimated for the Bay of Campeche continental shelf seep. In the Bay of Campeche's Cantarell oil field, the relative contributions of terrigenous, seep, upper-water primary production and anthropogenic carbon and nitrogen sources to the benthic food web are explored. Carbon isotopes in Yucatan shelf coral tissues are used to trace the natural and anthropogenic carbon sources affecting them. Finally, in the E Gulf of Mexico, the

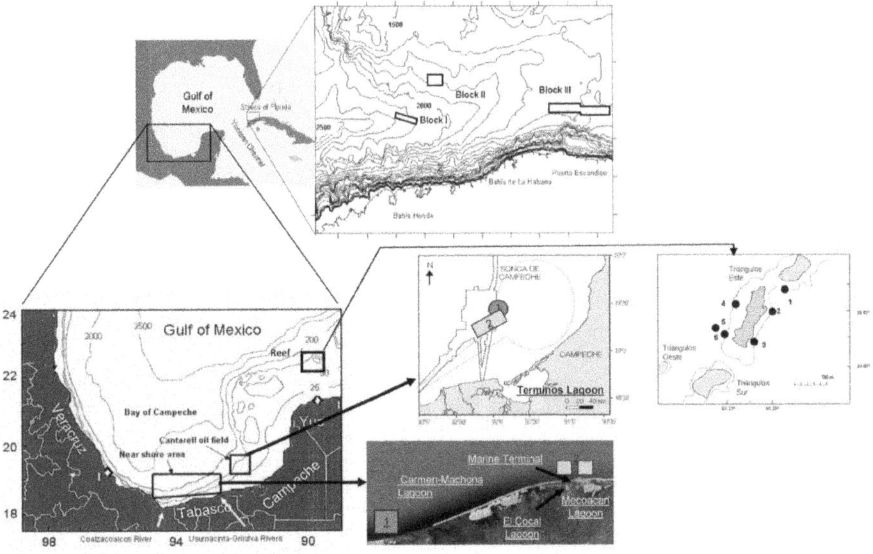

**Figure 1.**
*Southern and Eastern Gulf of Mexico study zones showing the deepwater sampling locations off NW Cuba slope region where sediment samples were collected (upper right panel), Bay of Campeche (middle and lower right panels) and Triangulos Reef in Yucatan Shelf (far right panel). Numbers in the Cantarell oil field region and near shore area depict the two cruises where particles, sediments and benthic organisms were collected. Numbers around Triangulos Reef are sampling sites. Depth contours in meters.*

long-term nitrogen sources to primary producers are evaluated using stable nitrogen isotopes of surface sediments.

## 2. Methodology

Stable carbon and nitrogen isotopes were analyzed in suspended particulate materials collected in the water columns, sediments, and animal matrices. Sample collection, prepping, and analysis for the four case studies are summarized in **Table 1**, details are given below.

### 2.1 Sampling

Suspended particles were collected in pre-combusted GF/F filters by filtering 2–10 L of seawater with a peristaltic pump, which retained particles with a nominal size of >7 μm. Samples were stored in petri dishes and frozen until analysis. Coastal and shelf benthic organisms and sediments were collected with a box corer or anchor dredge. Sediment cores (25 m long) from the E Gulf were retrieved with a Kullenberg piston corer. The recovered cores were sliced on board into 10 cm sections. Seven of these sections were analyzed in this study. The top 10 cm of the Bay of Campeche continental shelf and coastal sediment core samples were collected with a box corer. Corals and sponges were collected manually. Organisms and sediments were kept at freezing temperatures until analysis.

### 2.2 Sample treatment

In the laboratory, sediment and filtered particle samples were oven-dried, pulverized, and homogenized to a fine powder in a mortar. Sediment and reef

| Matrix | Method of collection | Sample treatment | Isotope analysis | Ancillary data |
| --- | --- | --- | --- | --- |
| Seep sediments ($^{13}$C) | Kullenberg corer | Drying, grinding, acidification, sieving | 1, 2 | Total hydrocarbons |
| Macrobenthic organisms ($^{13}$C, $^{15}$N) | Anchor-box dredge | Tissue extraction, rinsing, drying, grinding | 1, 2, 3 | — |
| Campeche surface sediments ($^{13}$C, $^{15}$N) | Box corer | Drying, grinding, acidification, sieving | 1, 2, 3 | Organic carbon, total nitrogen |
| Suspended seawater particles ($^{13}$C, $^{15}$N) | Seawater filtration | Drying, grinding | 1, 2, 4 | Organic carbon, total nitrogen |
| Sponge, algae, and coral tissue ($^{13}$C) | Collected manually | Drying, grinding, acidification, sieving | 1 | Polycyclic aromatic hydrocarbons |
| NW Cuba sediments ($^{13}$C, $^{15}$N) | Box corer | Drying, grinding, acidification, sieving | 1, 2 | Organic carbon, total nitrogen |

*1—Sealed tube combustion and a Finnigan MAT-252 stable isotope mass spectrometer (Laboratorio de Geoquímica del Petróleo, Instituto Mexicano del Petróleo). Reported values are averages of 10 replicate runs.*
*2—Continuous flow in a Europa ANCA-GSL elemental analyzer interfaced to a Europa 20–20 isotope ratio mass spectrometer (Rosenstiel School of Marine Sciences, University of Miami).*
*3—Sealed tube combustion and a Finnigan MAT-250 stable isotope mass spectrometer (Instituto de Geología, Universidad Nacional Autónoma de México). Reported values are averages of three replicate runs.*
*4—Continuous flow in a Costech ECS-4010 elemental analyzer interfaced to a Finnigan MAT-252 stable isotope ratio mass spectrometer (Department of Oceanography, Texas A&M University).*

**Table 1.**
*Summary of sample collection and treatment used to analyze stable carbon and nitrogen isotopes of the different sample matrices.*

organisms samples were acidified with 1 N HCl to eliminate the carbonate in the sediment matrix, washed with distilled water, and oven-dried at a temperature of <80°C. The previous studies have shown that acidification can change the $\delta^{15}$N values of the sample (e.g., [10]); however, with this concentration and type of acid, nitrogen isotopes are not significantly fractionated [11]. Organisms were oven-dried overnight at <60°C and then ground in a mortar into a fine powder. Samples analyzed by sealed tube combustion were mixed with copper oxide for carbon isotope analysis, or with elemental copper and copper oxide for simultaneous nitrogen and carbon isotope analyses, and sealed with a torch under vacuum in Pyrex or quartz tubes for carbon and nitrogen isotope analyses, respectively **(Table 1)**.

## 2.3 Sample analysis

Sealed tube samples were combusted at 550°C for carbon isotope analysis and at 900°C when both carbon and nitrogen isotopes were analyzed. The evolved combustion gases ($CO_2$ and $N_2$) were separated from the other combustion products by cryogenic distillation and analyzed in either a Finnigan MAT-250 or MAT-252 isotope-ratio mass spectrometer (IRMS). Samples analyzed by a continuous flow were placed inside crucibles in elemental analyzers interfaced to either a Europa 20–20 or a Finnigan MAT-252 IRMS. Runs were calibrated against standards of known isotopic composition and are reported relative to Pee Dee Belemnite (PDB) (or Vienna Pee Dee Belemnite (VPDB) for the Europa 20–20 samples) for carbon and air for nitrogen. The precision of the analyses for the four IRMS that were utilized ranged between ±0.09 and ±0.10‰ for $\delta^{13}$C and ±0.15 and ±0.26‰ for $\delta^{15}$N analyses.

## 2.4 Ancillary data

Seawater temperature and salinity were measured in situ with a Seabird CTD. Total petroleum hydrocarbons (TPH) in the Bay of Campeche seep and control core samples were analyzed by infrared spectrophotometry as per EPA Method 418.1 [12]. Organic carbon ($C_{org}$) and total nitrogen (TN) in sediments and $C_{org}$ and N in particles (particulate organic carbon (POC) and particulate nitrogen (PN)) were measured in elemental analyzers, where the carbon is quantified as $CO_2$ and nitrogen as $N_2$.

# 3. Results and discussion

## 3.1 Carbon apportionment in a bay of Campeche oil seep

Oil seeps are frequent in the S Gulf of Mexico both in the continental slopes [13] and in shallow waters of the Cantarell oil field [14], where major oil exploitation takes place. The volume of seeped oil in the region at any given time can be as high as 46 $m^3$ [14]. Satellite images of the region have shown that oil seepage between the years 2000 and 2002 appeared in nearly 80% of the analyzed images and area coverage ranged between 0.04 and 207 $km^2$ with an average of 32 $km^2$ [15].

The hydrocarbon concentrations and organic matter $\delta^{13}C$ distributions in a sediment core from a borehole drilled in a seep area near to Bay of Campeche's *Akal H* oil rig (92°20′W, 19°20′N) from the Cantarell oil field were compared with those from a nearby reference site (91°40′W, 20°20′N). The two sites lie at approximately 45 m water depth.

The depth distributions of TPH concentrations and $\delta^{13}C$ values in these sediments are given in **Figure 2**. TPH decreased linearly with a depth from 250 to 150 ppm at the reference site, which contrasts to concentrations at the seep site where values increased from 0.1% at 5 m depth to 35% at 20 m [16]. At both locations, $\delta^{13}C$ values showed similar profiles with maxima of −21.9‰ and −20.3‰ at 5 m depth decreasing to −26.5‰ and −22.0‰ at 20 m at the seep and reference sites, respectively.

The near-surface $\delta^{13}C$ values and TPH concentrations are concordant with a common carbon source in the sediment's upper layers of both sites. The topmost layer from the two cores suggests a mixture of organic matter from terrestrial and marine origins, because $\delta^{13}C$ values lie intermediate between values of marine algae (−19 to −16‰) and of $C_3$ land plants (−27‰) [17–19]. At the seep site, the decrease in the isotopic signature with depth and the concomitant increase of TPH are in clear concordance with a petrogenic hydrocarbon origin in the deep section of the core, in line with the isotopic range of Campeche Sound's oil families [20, 21]. In turn, the $\delta^{13}C$ signature at the reference site agrees with the expected sedimentary organic matter depth distribution from several Gulf of Mexico sites [17, 22].

The relative $\delta^{13}C$ maxima (−22‰ at the reference site and −23‰ at the seep) appear at around 5 m core depth. Below, the $\delta^{13}C$ signal shows that downcore sediments are $^{13}C$-depleted to the maximum sampled depth. Several authors have recorded isotopic shifts in the Gulf of Mexico's upper layers of sediment cores [18, 19, 22]. It is generally accepted that downcore $\delta^{13}C$ depletions reflect a terrestrial plant input resulting from a lowering of sea level attributable to the Last Glacial Maximum of the Late Pleistocene (ca. 25,000 B.P.). The upper sediments of different regions of the Gulf of Mexico show $\delta^{13}C$ values on the order of −22 to −18‰ which overlay a thick layer of isotopically lighter values (−23 to −27‰) [17, 23]. In turn, shifts to enriched $\delta^{13}C$ values occur during interglacial periods. The

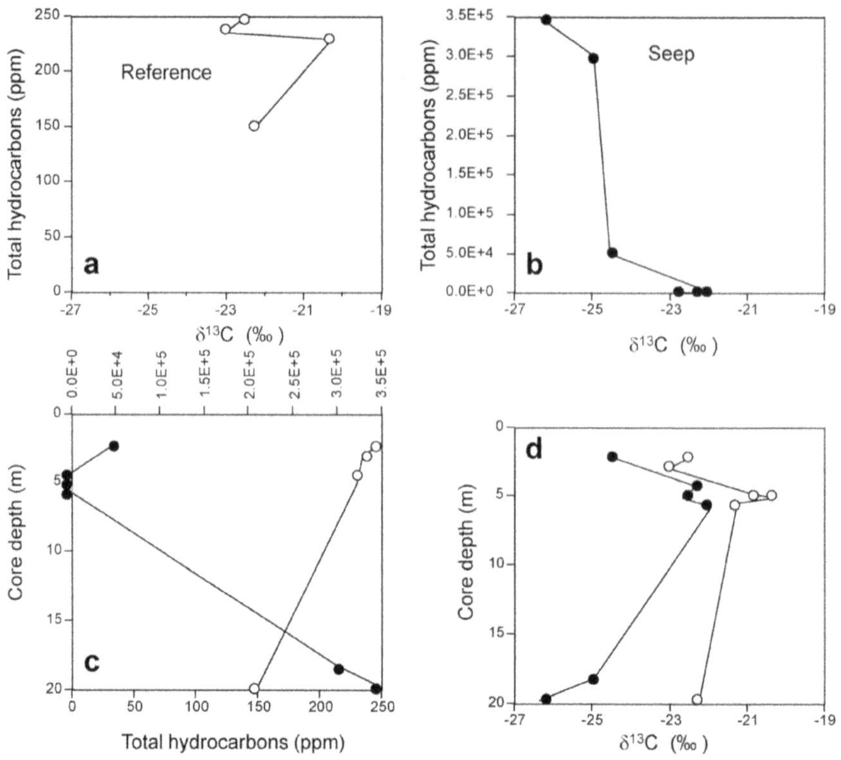

**Figure 2.**
*Upper panel: Total petroleum hydrocarbon concentration vs. $\delta^{13}C$ composition for (a) reference (opened circles) and (b) seep (closed circles) sites from Bay of Campeche seep region. Lower panel: Depth distribution of (c) total hydrocarbon concentration and (d) $\delta^{13}C$ values of the reference and seep sites.*

magnitudes of these isotopic shifts vary in intensity depending on their location within the Gulf and on prevailing upper-water circulation patterns. Off the Brazos-Trinity Basin and in Pigmy Basin, in the northern Gulf, the isotopic signal varies from −27 to −20‰ as a result of the dominance of terrigenous organic matter during the lowering of sea level [18, 19, 22]. In contrast, this isotopic signal shift is not evident at Ursa Basin, off the Mississippi River delta, where the continuous terrestrial drain overwhelms the isotopically heavier marine signal.

Considering that in North America the Last Glacial Maximum ended approximately at 12.5 ky [24], and that a $\delta^{13}C$ maximum was found at 5 m depth, then, using Eq. (3), an average sedimentation rate of 40 cm/ky can roughly be estimated from the depth of this isotopic shift. As shown in **Figure 3**, Gulf of Mexico sedimentation rates vary considerably, from 4 to 11 m/ky off the Mississippi River fan [25, 26], 20–40 cm/ky off Louisiana continental slope, and deep western Gulf of Mexico [26, 27] to 5 cm/ky off NE Gulf's continental shelf [28]. Because Bay of Campeche's continental shelf is near the discharge area of the Coatzacoalcos and the Grijalva-Usumacinta River systems, and the continental shelf is overlain by surface waters of high productivity [29, 30], the calculated sedimentation rate of 40 cm/ky is reasonable and is similar to those estimated for the Gulf of Mexico deep western and northern slope environments [26, 27].

From the $\delta^{13}C$ values of the isotopic maximum depth from the two sites, the relative contribution of seep-derived petrogenic hydrocarbons at that depth can be estimated using the stable isotope-mass balance equation (Eq. (3)). Assuming that the reference site's $\delta^{13}C$ maximum value (−20.3‰) represents the isotopic

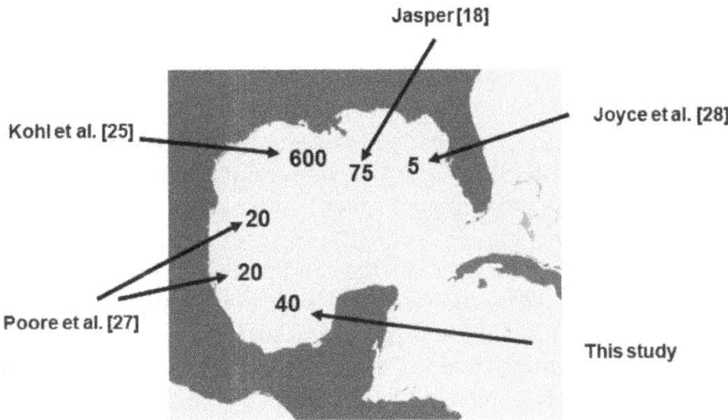


**Figure 3.**
*Estimated sedimentation rate for Bank of Campeche. Also shown are sedimentation rates for several locations of the Gulf of Mexico. Data in cm/ky.*

composition of biogenic carbon contribution, and that the $\delta^{13}$C value at 20 m depth of the seep site ($-26.4‰$) is entirely petrogenic, then the isotopic composition at the isotopic maximum depth of the seep site ($-21.94‰$) is constituted by approximately 27.6% biogenic carbon.

Summarizing, the carbon isotope depth profiles were used to distinguish a biogenic hydrocarbon source in the upper layers of the two cores and oil-derived TPH in the deeper sections of the seep site. The carbon isotope distribution with depth also identified the depth of a major paleoclimatic event in the Bay of Campeche, which precluded the input of terrestrial organic matter into the adjacent shelf. Considering the distinct carbon isotope compositions for the two principal carbon sources in these sediments, the oil-derived carbon contribution to the upper sediments of the seep area is estimated at around 72.4%.

## 3.2 Carbon and nitrogen sources to the benthic food web in the bay of Campeche oil field region

Potential carbon sources to the Bay of Campeche continental shelf and coastal system include in situ primary production, upwelling, terrigenous inputs from rivers, natural seeps, and offshore oil production. In order to discern the carbon and nitrogen sources and infer the benthic trophic structures in the coastal zone and Bay of Campeche's Cantarell oil field region, sediments, suspended particles, and macrobenthic organisms were sampled and analyzed for stable carbon and nitrogen isotopes.

In situ primary production and resulting particle flux can be important sources of organic matter to continental shelf and pelagic sediments. Near-surface water chlorophyll concentrations are relatively high over the Bay of Campeche shelf and decrease offshore [29–31]. According to Hidalgo-González and Alvarez-Borrego [31], the average surface chlorophyll-*a* concentration for the S Gulf of Mexico is 28 mg/ m$^3$, which is more than 2.5 times higher than at the Gulf's open waters. The yearlong upcoast current flowing along this section of the shelf partially explains this increment, because this flow induces upwelling of nutrient-rich subsurface waters [30]. Additionally, the presence of cyclonic rings in the Bay of Campeche resulting from the topographic effect of Campeche Canyon [32] enhances vertical water mass movements which are favorable to primary producers.

Continental drainage is another potential source of organic matter in S Gulf of Mexico. The principal discharge to the region is the Grijalva-Usumacinta River system (the second most important river discharge to the Gulf of Mexico after the Mississippi River), followed by the Coatzacoalcos River. Their flows are seasonally variable with a five- to tenfold oscillation between the highest and lowest runoffs. The presence of drowned land, soil drainage from rainforests, and a series of coastal lagoons, such as the Carmen-Machona Lagoon, underscores the importance of terrestrially derived organic matter flow to the adjacent continental shelf. This input of terrestrial organic matter favors reducing conditions in the deltas and inner shelf sediments of the Coatzacoalcos and Grijalva-Usumacinta Rivers, and thus organic matter tends to be preserved [33]. Off the Coatzacoalcos River, the seasonal convergence of the western downcoast and eastern upcoast coastal currents in the region disperses the supply of terrestrial organic matter to the region [30]. Strong haline and thermal gradients develop off the Coatzacoalcos and Grijalva-Usumacinta River's delta which transport particulate matter from the inner shelf to the shelf break [29].

Another potential organic carbon contributor to the Bay of Campeche is offshore oil exploration and production. Off the Tabasco coast, the Dos Bocas Marine Terminal concentrates the oil produced offshore which in turn is shipped or distributed inland. The produced water (that which is separated from the extracted oil) is partly discharged from the marine terminal to the coastal zone by a submarine diffuser. This partially treated discharged water is usually high in metals and carbon products. Oily and graywater discharges from offshore oil rigs can contribute with organic carbon and nitrogenous nutrients, respectively, to the adjacent environment.

The study zone is localized in the transition zone between the carbonate and terrigenous sediments of the Bay of Campeche continental shelf [34] and spans from nearshore sampling sites off the Coatzacoalcos and Grijalva-Usumacinta Rivers to the 60 m depth isobath in the oil rig zone (**Figure 1**). Results from two cruises to the sampled area are used here to infer the relative importance of the different carbon and nitrogen sources to benthic organisms. In the first cruise, coastal benthic organisms and sediments were collected at three stations off Coatzacoalcos River and Grijalva-Usumacinta deltas and at two stations off Carmen-Machona Lagoon [35]. Samples were also collected 12 nmi due WNW, and 1 nmi off Cocal Lagoon, which is connected to the Carmen-Machona Lagoon System. These samples are used as a coastal reference region. Additionally, continental shelf benthic organisms and sediments were sampled at five locations outside the Cantarell oil rig field at water depths between 38 and 44 m; these samples are used as an offshore reference region.

In a second cruise, sediments and suspended particles of the coastal zone were sampled 1.2 nmi north of the Dos Bocas Marine Terminal discharge zone, at a coastal reference site, inside the Cantarell oil rig field and at two offshore reference regions. Samples were collected at different distances from the discharge points of the marine terminal and oil rigs. **Table 2** summarizes the type of samples collected in each zone.

**Figure 4** shows the percentile distribution as boxplots of the $\delta^{13}C$ and $\delta^{15}N$ values in sediments and suspended particles of the coastal and offshore regions. Also shown are the isotopic compositions of potential sources such as particles in oily and graywaters discharged from the oil rigs, in produced water discharged by the Marine Terminal, and the $\delta^{13}C$ average value of Cantarell crude ($-27.9‰$). The average $\delta^{13}C$ values of oily and produced waters ($-27.2$ and $-26.7‰$, respectively) are similar to that of Cantarell oil, because they have a common hydrocarbon source.

| Region | Suspended particles | Sediments | Macrobenthos |
|---|---|---|---|
| Coastal reference | — | 11 | 6 |
| Marine terminal | 33 | 9 | — |
| Offshore reference | 19 | 9 | 7 |
| Oil rigs | 51 | 14 | — |

**Table 2.**
*Number of samples collected in the Bay of Campeche coastal, offshore, and seep regions.*

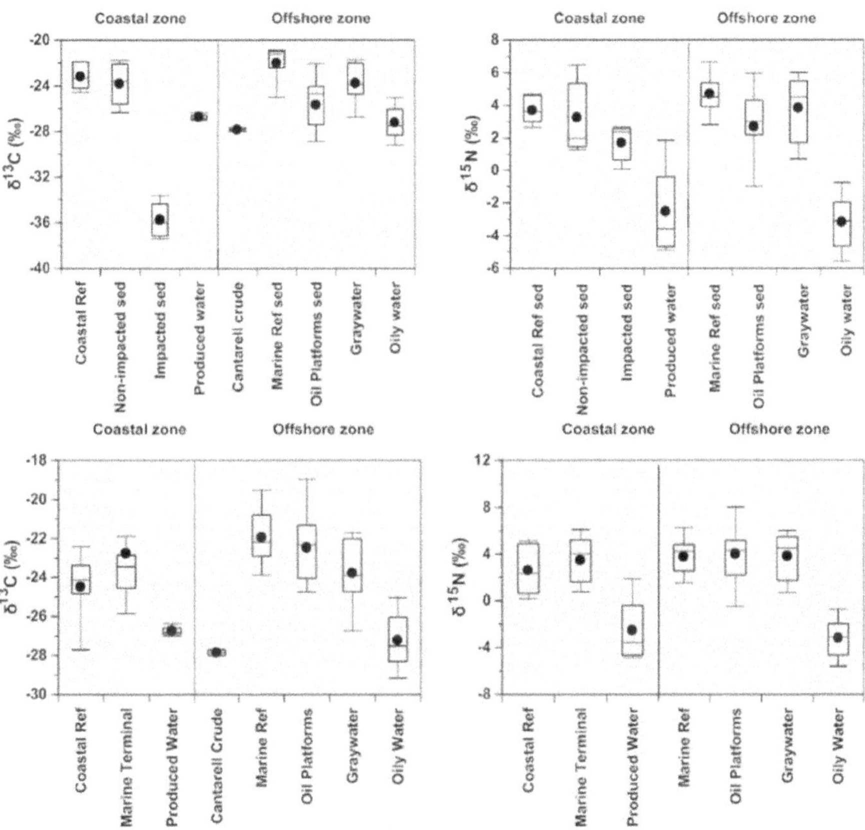

**Figure 4.**
*Box charts of carbon and nitrogen isotopes of marine terminal and costal reference sediments (upper panel) and suspended particles (lower panel) in the coastal zone and marine platform, and reference sediments in the offshore zone. Also shown are the $\delta^{13}C$ and $\delta^{15}N$ values of particles in produced water to the coastal zone, and of oily and grey waters in the offshore zone. The average $\delta^{13}C$ value of Cantarell oil [19] is also depicted. Endmembers denote the 10 and 90 percentiles, and horizontal lines of the box denote 25, 50 and 75% quartiles. Symbol denotes average value.*

Because Marine Terminal sediments spanned a wide range of $\delta^{13}C$ values, these were further separated into impacted $^{13}C$-depleted sediments (average $\delta^{13}C$ of $-35.7‰$) and non-impacted sediments whose average $\delta^{13}C$ value ($-23.8‰$) was similar to the average of the coastal reference zone ($-23.3‰$). The average $\delta^{13}C$ value of impacted sediments was much lighter than Cantarell oil and oily waters and suggests isotopic fractionation by microbial activity metabolizing carbon discharged in produced waters [36]. The nitrogen isotopic composition of coastal sediments averaged 3.7‰, statistically similar with that of non-impacted sediments. In contrast, the boxplot of the nitrogen isotope composition of impacted sediments is lighter than that of coastal and non-impacted sediments and suggests a contribution of the lighter $\delta^{15}N$ values from the produced water.

The average carbon isotope composition of offshore sediments ranged from −22.0‰ in the reference zone to −25.7‰ in the oil rig zone, which are heavier than the $\delta^{13}C$ value of Cantarell crude and of the particles discharged with the oily water. The wide spread of $\delta^{13}C$ values for oil rig sediments suggests the input of in situ primary production and hydrocarbons in different proportions. In turn $\delta^{15}N$ values of sedimentary nitrogen mostly ranged between 2 and 5‰, in contrast to the more [15]N-depleted discharge of oily waters.

In contrast to the sediments, the $\delta^{13}C$ and $\delta^{15}N$ signatures of suspended particles were statistically similar in the three regions, mostly due to the wide range between maximum and minimum values (**Table 3**). However, histograms of $\delta^{13}C$ values for POC and PN show that the modal $\delta^{13}C$ values in the coastal zone plus marine terminal were generally 1–2‰ lighter than in the Cantarell region (modal values of −25 to −23‰ vs. −23 to −21‰; **Figure 5**). In turn, the distribution of $\delta^{15}N$ values was similar at the two sites, with modal values of $\delta^{15}N$ which were centered at 2 and 5‰, suggesting two principal common nitrogen sources to primary producers in these regions.

The input and fate of terrestrial organic matter into the coastal zone across the coastal plume's salinity gradient at the eastern and western flanks of the oil marine terminal and a river mouth are shown in **Figure 6**. High POC concentrations were

| Matrix | Region/Discharge | $\delta^{13}C$ (‰) | | $\delta^{15}N$ (‰) | |
|---|---|---|---|---|---|
| | | Average | Range | Average | Range |
| *Coastal sediments* | River outflow | −23.22 (1.23) [a] | −24.60 to −21.90 | 3.68 (0.84) [c] | 2.60 to 4.70 |
| | Coastal non-impacted | −23.84 (1.98) [b] | −25.82 to −21.74 | 3.83 (2.29) | 2.36 to 6.47 |
| | Coastal impacted | −35.75 (1.72) [a,b] | −37.40 to −33.64 | 1.11 (0.97) [c] | 0.06 to 1.97 |
| *Offshore sediments* | Offshore reference | −22.01 (1.83) [d] | −25.14 to −20.86 | 4.69 (1.41) [e] | 2.50 to 7.40 |
| | Cantarell oil field | −25.69 (3.12) [d] | −33.87 to −21.10 | 2.66 (2.63) [e] | −3.17 to 6.22 |
| *Coastal particles* | Produced water | −26.74 (0.24) [f] | −27.01 to −26.33 | −2.26 (2.75) [g,h] | −4.90 to 1.87 |
| | Oily discharge | −27.24 (1.26) [i,j,k] | −29.51 to −24.66 | −1.35 (3.39) [l,m,n] | −5.63 to 5.82 |
| | Graywater discharge | −24.27 (3.97) [i] | −37.66 to −21.25 | 3.83 (2.35) [g,l,o] | 0.43 to 8.40 |
| | Coastal zone | −23.74 (1.84) [j] | −29.48 to −21.78 | 2.36 (2.18) [m,o] | −1.24 to 5.21 |
| | Marine terminal | −23.92 (1.43) [f, k] | −27.40 to −21.74 | 3.52 (1.97) [h,n] | 0.38 to 6.47 |
| *Offshore particles* | Offshore reference | −21.89 (1.83) | −26.50 to −18.15 | 3.77 (1.87) | −0.96 to 2.60 |
| | Cantarell oil field | −22.46 (2.61) | −30.38 to −16.33 | 4.01 (3.32) | −3.17 to 14.82 |

*Superscripts b, i, and o denote significant differences between regions or discharge values (t test, $p < 0.05$). Superscripts a, b, c, f, g, h j, k, l, m, and n denote highly significant differences between regions and/or discharge values (t test, $p < 0.01$).*

**Table 3.**
*Basic statistics for stable carbon ($\delta^{13}C$) and nitrogen ($\delta^{15}N$) isotope values of surface sediments and suspended particles in the river's outflow, coastal zone, oil marine terminal, Cantarell oil field, and reference offshore region. Standard deviations are given in parentheses.*

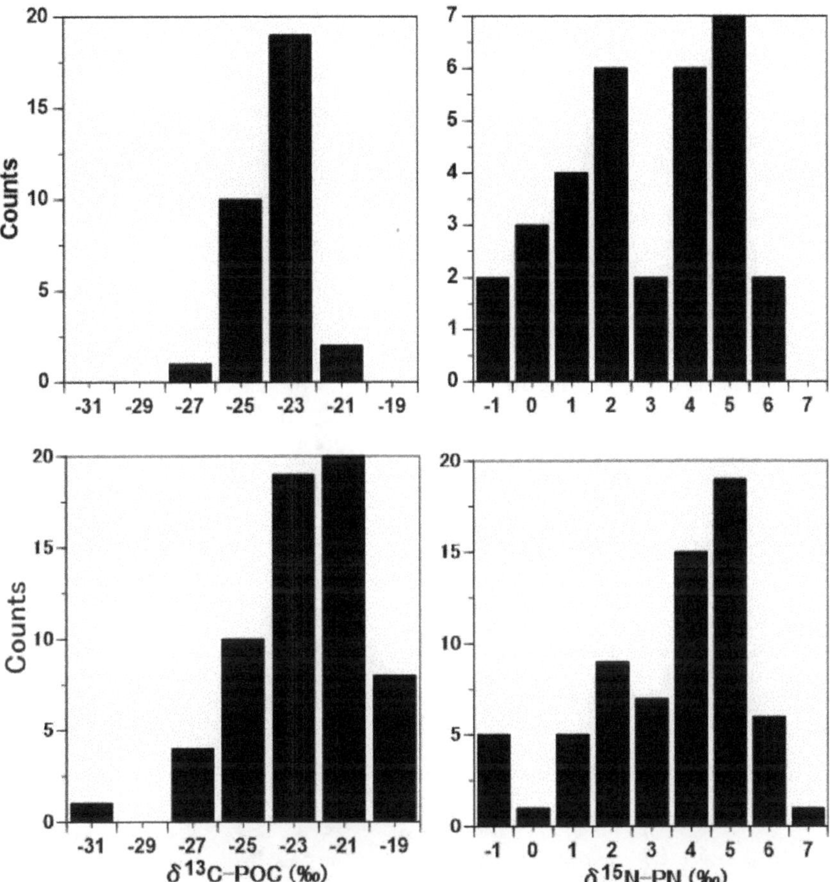

**Figure 5.**
*Histograms of the isotopic composition of particulate organic carbon ($\delta^{13}$C-POC) and particulate nitrogen ($\delta^{15}$N-PN) for the coastal region (upper panel) and Cantarell oil field region (lower panel).*

associated with low salinity waters and decreased across the salinity gradient to typically lower ocean values, indicating that the relatively high POC concentrations of continental origin decrease as the plume is diluted across the salinity plume. In contrast, the peak PN concentration at a salinity of 32 suggests a maximum N uptake at optimal inorganic nitrogen and light conditions. Likewise, the low $\delta^{13}$C values of terrestrial particles are concordantly diluted as POC concentrations decrease. Terrestrially derived particles have high C:N molar ratios and low $\delta^{13}$C values, while marine particles approach C:N molar ratios of 6:1 and $\delta^{13}$C values of approx. −23‰. For example, $\delta^{13}$C measurements off the mouth of the Coatzacoalcos River yielded values of −28.5‰ which changed to −23.0‰, indicating a rapid dilution of terrestrially derived organic matter near the river's mouth [37]. These changes have been recorded in other regions of the Gulf of Mexico, such as the Mississippi River plume where both dissolved species [38] and particulate species are diluted [39].

Scatterplots for coastal and Cantarell oil area sediments, benthic organisms, and suspended particles are summarized in **Figure** 7. Also shown is the stable isotope composition of oil from the Cantarell area [20, 21]. Most sedimentary $\delta^{13}$C data points for coastal and shelf regions fall around −22‰ and $\delta^{15}$N values between 3 and 7‰. However, a few coastal samples and one shelf sediment appear to incorporate biogenic methane carbon as suggested by the depleted $\delta^{13}$C values

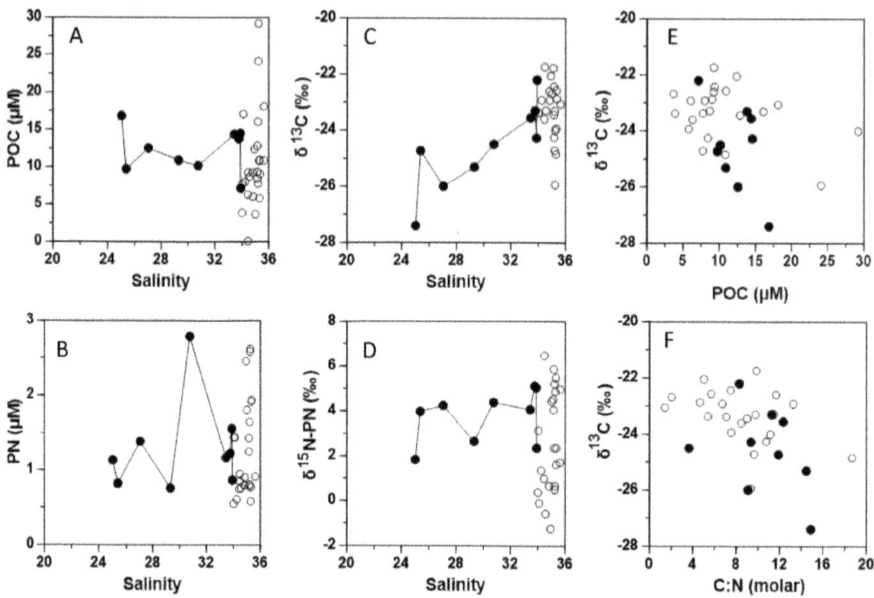

**Figure 6.**
*Salinity vs. parameter scatterplots showing (A) POC, (B) PN, (C) $\delta^{13}C$, and (D) $\delta^{15}N$ distributions along the salinity gradient, and (E) $\delta^{13}C$ vs. POC and (F) $\delta^{13}C$ vs. C:N molar rations of particulate matter in Bay of Campeche coastal zone. Open circles are high salinity samples, closed symbols are low salinity (<34) samples.*

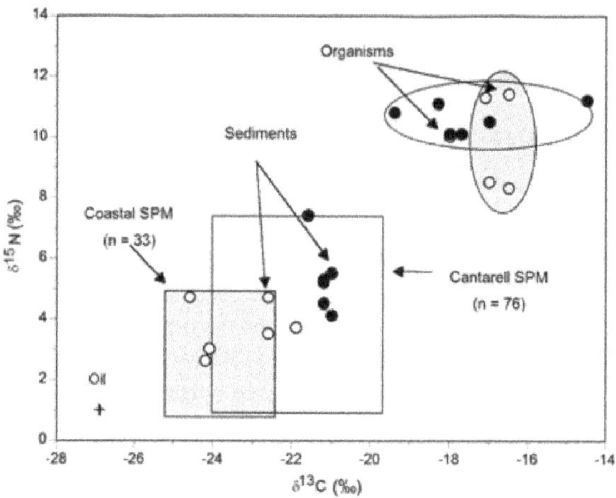


**Figure 7.**
*$\delta^{15}N$ vs. $\delta^{13}C$ scatterplots of coastal (open circles) and shelf (closed circles) particles (SPM), sediments and benthic organisms of Campeche shelf. The squares enclose the range of isotopic values of suspended particles (not shown) and the cross denotes the isotopic composition of Cantarell oil [20, 21].*

between −38 and −34‰. By contrast, the carbon isotope composition of coastal particulate matter ranged between −26 and −22‰. Suspended particles in the shelf area spanned a greater range of $\delta^{13}C$ and $\delta^{15}N$ values (−26 to −14‰ and −4 to 10‰, respectively). Considering that most particulate matter is composed of autotrophic organisms, it appears that coastal area and shelf area were constituted by different phytoplankton assemblages. At Dos Bocas Marine Terminal area, the terrestrially derived organic carbon is isotopically lighter than that of the coastal region. POC concentrations decreased along the salinity gradient and increased at higher salinities. Nitrate concentration decreased somewhat while PN remained constant along the salinity gradient with oscillating isotopic compositions. As shown in the figure, the superposition of carbon and nitrogen isotopes of the water column-suspended particulate matter from nearshore and Cantarell areas with the isotopic values of sediments suggests that the predominant source of sedimentary $C_{org}$ and TN in the Cantarell area is the downward flux of upper-layer production. However, the most negative value ($\delta^{13}C$ of $-38\permil$) of some coastal and Cantarell sediment samples suggests the presence of hydrocarbons, particularly off the marine terminal. By contrast, the isotopically lighter carbon in nearshore area particles and sediments reflects a mixture of marine organic matter with that of terrestrially derived origin, principally from the Grijalva-Usumacinta River, which is the most important contributor of terrestrial organic matter to the southern Gulf of Mexico. The predominant coastal current in this sector of the shelf flows westward [30], and, along the coastal plume, numerous smaller discharges would supply additional organic matter of terrestrial origin.

Based on stable isotope analysis, no evidence of seep- or oil-derived carbon in the sampled sediments of the coastal zone and Cantarell area were found. The measured sediment carbon isotope values are also more $^{13}C$-enriched than those found at other seep areas, such as in the NW Gulf of Mexico (e.g., $-27.5$ to $-26.5\permil$ [36]) or gas hydrates from southern Gulf of Mexico slope cold seeps [13]. Likewise, $\delta^{13}C$ values for surface sediments sampled in offshore oil rigs of the Cantarell area are significantly lighter ($-27.5$ to $-26.4\permil$ [40]) than those found in this study.

Sampled macrobenthic organisms in the coastal region consisted mainly of penaeid shrimp, and those from Cantarell included infaunal (polychaetes and sipunculids) and epifaunal (ascidians, echinoderms, and decapods) species. In the two regions, the isotopic compositions of organisms were heavier in carbon and nitrogen than those from sediments and corresponding suspended particles, with $\delta^{13}C$ values oscillating between $-19.4$ and $-17.0\permil$ and $\delta^{15}N$ between 8 and $11\permil$. Considering that on average there is an increment of $3.5\permil$ in $\delta^{15}N$ between one trophic level and another [2] and around $1.5\permil$ in $\delta^{13}C$ [1], then the Cantarell area-sampled specimens are about two trophic levels above primary producers. In the coastal region, the data suggest that the sampled benthic species are two and three trophic levels above producers. These results are in line with a phytoplankton-based benthic trophic chain [41]. In contrast, seep communities derive their carbon sources from isotopically light carbon from oil or gas through chemosynthetic microorganisms, and resulting carbon and nitrogen isotopic values are much lighter, on the order of $-60$ to $-30\permil$ for $\delta^{13}C$ and $-17$ to $2.5\permil$ for $\delta^{15}N$ [39, 41].

In summary, the contribution of seep-derived carbon is negligible on a shelf-wide scale for this region. The principal carbon source to benthic organisms in this continental shelf is water column production, with a terrestrial contribution in nearshore waters, which is rapidly diluted as the coastal plume is diluted. It follows that organic carbon derived from the shelf's high primary production overwhelms that from oil seeps and from offshore drilling in this region. However, the sedimentary carbon isotopic composition of the oil terminal suggests the local presence of hydrocarbon-derived organic carbon.

Although stable isotopes are useful tracers of the source and flow of biogeochemically relevant elements in trophic web studies, fractionation processes of nitrogen [2] and even carbon [1, 42], along with mixed diets, may yield an incorrect interpretation of food sources. This problem may be solved by including a third stable isotope (e.g., $^{34}S/^{32}S$, $^{18}O/^{16}O$) which complement the other two [42]. For example, the complementary use of $\delta^{13}C$, $\delta^{15}N$, and $\delta^{34}S$ was used to elucidate the complex trophic web of deep-sea hydrothermal vent systems in the southern Gulf of California, where different carbon sources and assimilation processes along with different sulfide sources (magmatic, biogenic, and photosynthetic) co-occur [43].

### 3.3 Organic carbon and nitrogen sources to NW Cuba deep-sea sediments

In the E Gulf of Mexico, Cuba's NW sector (**Figure 1**), localized between the Yucatan Channel and the Straits of Florida, is a highly dynamic region. In this region the inflowing Yucatan Current brings in water into the Gulf of Mexico, and the Florida Current transports water from the Gulf out into the Atlantic Ocean via the Gulf Stream. Across the channel near-surface current velocities extend throughout the water column. Low nutrient concentrations in part explain the low surface chlorophyll concentrations reported for the E Gulf of Mexico and Yucatan Channel [44]. As a consequence, atmospheric nitrogen, with a $\delta^{15}N$ near 0‰, is an important potential nitrogen source in this part of the Gulf of Mexico [45]. There is a relatively scarce contribution of terrigenous organic matter input when compared to the N and S Gulf of Mexico, where the Mississippi and the Grijalva-Usumacinta Rivers drain, respectively. Therefore, sediments are dominated by calcareous oozes and marls, although surface sediment magnetic susceptibility distribution for the Gulf of Mexico suggests detrital sediments originating from igneous and metamorphic soils from Cuba depositing off the island's NW continental slope and abyssal plain [46].

Recent search for deep-sea fossil fuels and gas hydrates has renewed interest in the study of deep-sea processes in the Gulf of Mexico [47]. In 2002 a multidisciplinary research group explored the seabed off the northwestern coast of Cuba with the purpose of detecting potential deep-sea hydrocarbon seeping areas [47, 48]. Stable carbon and nitrogen isotopes in surface sediments are used here to detect possible seeping in the area and to infer the relative importance of nitrogen fixation as a source of sedimentary organic matter in the Cuban northwest slope.

During the period of study, upper-water currents in the study zone were on the order of 180 cm/s with an ESE direction, and near-bottom currents reached 40 cm/s (**Figure 8**). The thermocline, which separates the well-mixed upper layer from the stratified deeper waters, was localized at around 75–100 m depth, precluding the vertical advection of subsurface nutrients to upper waters.

Three blocks located in Cuba's exclusive economic zone were explored off the northern slope of Cuba (**Figure 1**). Three samples were collected in Block I, the westernmost site localized 45 km north of the island at a depth of nearly 2200 m. Block II laid at 1800 m depth at 50 km north of the island, and samples were taken at three sites. The easternmost Block III had an elongated surface localized at a depth of 1500 m where six sites were collected. In total, 12 surface samples were subsampled from box-core deployments for carbon and nitrogen stable isotopes and $C_{org}$ and TN analyses.

**Figure 9** gives the scatterplots for $\delta^{13}C$, $\delta^{15}N$, $C_{org}$, TN, and C:N ratios for surface samples from NW Cuba continental slope. $C_{org}$ and TN concentrations were generally low and varied between 0.2 and 0.8% and 0.06 and 0.15%, respectively. Stable isotope values of the organic carbon fraction ranged between −19.1 and −18.5‰, and most $\delta^{15}N$ values varied between 5.4 and 6.4‰ [49]. An analysis of variance for surface $C_{org}$, TN, and isotope values showed no statistical difference between sites; however, molar C:N ratios were significantly higher at Block II relative to Block III (**Table 4**).

The study zone's $C_{org}$ and TN concentrations are nearly tenfold lower of surface sediments than the southern and northwest Gulf of Mexico [49, 50]. Several factors appear to determine these low $C_{org}$ and TN concentrations. A fraction of water entering through the Yucatan Channel from the Caribbean Sea deviates eastward

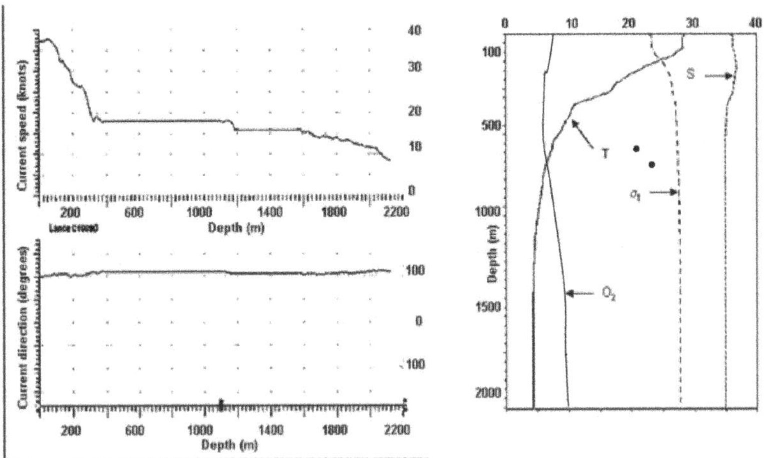

**Figure 8.**
*Representative oceanographic conditions for NW slope of Cuba waters during July 2002. Upper left panel: current velocity (knots) vs. depth (m). Lower left panel: current direction (degrees) vs. depth. Right panel; vertical profiles of temperatures (T, °C), salinity (S, per mile), sigma-t ($\sigma_t$, (1-density) × 1000) and dissolved oxygen ($O_2$, ml/L).*

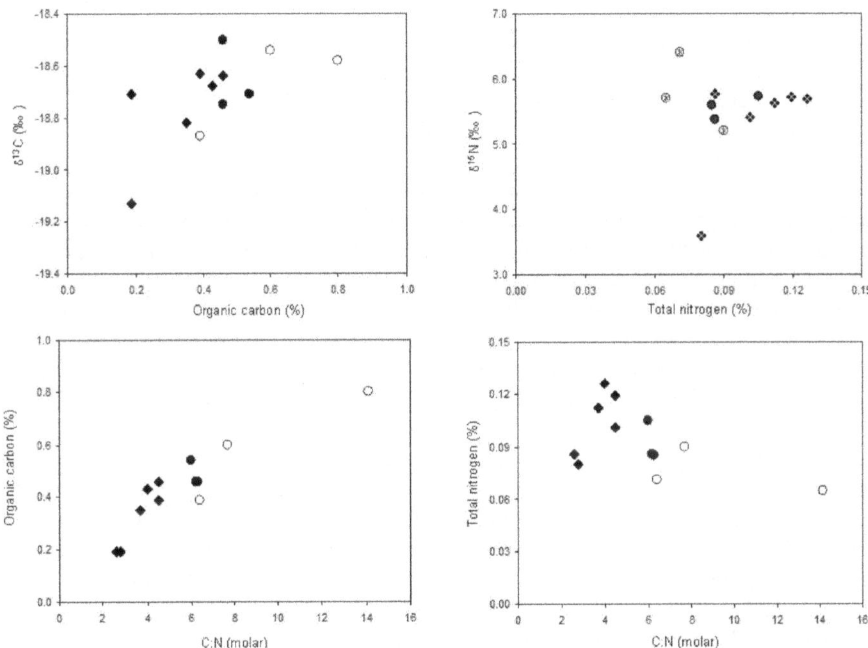

**Figure 9.**
*Scatterplots of: (A) $\delta^{13}C$ vs. organic carbon concentration, (B) $\delta^{15}N$ vs. total nitrogen concentration, (C) organic carbon concentration vs. C:N molar ratios, and (D) total nitrogen concentration vs. C:N molar ratios, in surface sediments from Block I (open circles), Block II (closed circles) and Block III (diamonds). See **Figure 1** for site locations.*

upon entering the Gulf of Mexico, and, because the flow accelerates as it transits through the channel, the thermocline deepens in the eastern side of the current [51], advecting nutrient-impoverished upper waters into the eastern Gulf [44]. Additionally, the northwest side of Cuba does not show a significant input of terrigenous

| Variable | SS$_{effect}$[a] | MS$_{effect}$[b] | M$_{error}$[c] | F[d] | p* |
|---|---|---|---|---|---|
| % C$_{org}$ | 0.1463 | 0.0731 | 0.0176 | 4.1590 | 0.0526 |
| % TN | 0.0017 | 0.0008 | 0.0003 | 3.2691 | 0.0857 |
| C:N (molar) | 66.1112 | 33.0558 | 4.1572 | 7.9514 | 0.0103* |
| $\delta^{13}$C (‰) | 0.0365 | 0.0182 | 0.0312 | 0.5833 | 0.5578 |
| $\delta^{15}$N (‰) | 0.4690 | 0.2345 | 0.4821 | 0.4864 | 0.6301 |

[a]*Sum of squares between sites (degrees of freedom: 2).*
[b]*Mean square between sites.*
[c]*Mean square of error (degrees of freedom: 9).*
[d]*Fisher test.*
**Significant difference (p < 0.05).*

**Table 4.**
*One-way analysis of variance for organic carbon, total nitrogen, molar C:N ratio, $\delta^{13}$C, and $\delta^{15}$N in surface sediments from Block I (three samples), Block II (three samples), and Block III (six samples) of northwest slope of Cuba.*

material through continental drainage, when compared to the riverine inputs from the northern and southern gulfs. In these nutrient-impoverished surface waters, nitrogen-fixing cyanobacteria have a competitive advantage, and *Trichodesmium* sp. blooms are frequent [44, 51].

The carbon and nitrogen isotopic composition of surface sedimentary organic matter would be similar to that of the overlying euphotic zone if the transit time of particulate organic matter from upper waters to the seafloor is short. In contrast, if particulate organic matter has a longer residence time in the water column, then in situ processes, such as microbial degradation, would transform the isotopic composition acquired in the surface waters by the selective removal of carbon and nitrogen compounds [19, 52]. In this study, the average $\delta^{13}$C value of $-18.7$‰ in sediments is similar to previously published values for surface water particles from the eastern Gulf of Mexico [35, 44], which suggests a rapid transfer or organic carbon from the upper waters to the sediments. Sedimentary $\delta^{13}$C values further suggest that there was no evidence of extensive hydrocarbon seeping in the studied samples.

Under oligotrophic conditions nutrient limitation in the photic zone precludes nitrogen isotope fractionation, and plankton acquires the isotopic composition of the available nitrate diffusing from below the thermocline [53, 54]. In northwest Gulf of Mexico slope waters, the $\delta^{15}$N composition of nitrate ($\delta^{15}$N$_{average}$ = 5.0‰) and mixed-layer PN (5.5‰) suggests that nitrate diffusing from the thermocline into the photic zone is an important nitrogen source for phytoplankton in that region [39]. If organic matter at the northwest Cuba slope is transferred rapidly from the euphotic zone to the sediments, then the isotopic composition of sedimentary nitrogen ($\delta^{15}$N$_{average}$ = 5.5‰) would be similar to that of upper-water PN. The measured sedimentary nitrogen isotope composition suggests that the principal, long-term nitrogen source to the sediments off the northwest slope of Cuba is nitrate diffusing from the thermocline, which is then deposited in the sediments by large particles with a short residence time in the water column. Surface sediment C:N molar ratios further suggest a predominantly marine origin for organic matter of NW slope of Cuba. Marine-derived organic matter has molar C:N ratios between 4 and 10, which contrast with terrestrially derived matter with ratios above 20, mostly resulting from the abundance of cellulose in its structure [50]. The significantly lower ratios at Block III relative to Block II result from the lowest C$_{org}$ concentrations and highest TN values in the former site which is nearer to the coastline.

### 3.4 Oil-related baseline levels of a Bank of Campeche coral reef

The high degree of structural complexity and species interdependence renders coral reefs as highly vulnerable ecosystems to natural or anthropogenically induced changes. For example, high mortality has been observed for coral larvae by water-accommodated fraction of fuel oil, dispersed oil, and oil dispersant at concentration levels with an order of magnitude lower than expected concentrations from an oil slick [55].

One of the three reef systems of the Mexican Gulf of Mexico is located in Campeche Bank, off the northwest edge of Yucatan shelf (**Figure 1**). The system is constituted by the emerged reefs Arrecife Alacranes, Cayo Arenas, Cayo Arcas, and Triángulos and by the submerged banks Banco Ingleses and Bajo Obispos [56]. Most of these reefs and banks cover a small area ($<20$ km$^2$), with the notable exception of Alacranes which is over 30-fold bigger [56].

Stable carbon isotopes and polycyclic aromatic hydrocarbons (PAHs) were measured in coral, sponge, and algae tissue samples from Triángulos Reef off Campeche Bank (NW of the Yucatan Peninsula) collected in September 2001 at bottom depths ranging between 8 and 19 m. The reef is adjacent to the Bay of Campeche where 80% of the country's crude oil is extracted and a nearby offshore terminal loads petroleum to oil tankers at Cayo Arcas [57]. The purpose of the study was therefore to determine if anthropogenic activity from the nearby offshore oil terminal is detected in this reef system. A second objective was to provide baseline stable isotope and PAH data for the reef systems of that region. Six sponges and six corals (*Montastraea cavernosa*) and two benthic algae samples were collected in the eastern and western flanks of Triángulos Reef at water depths varying between 8 and 18 m. Because the zooxanthellae were not separated from the coral, results represent a mixture of the host tissue and its symbiotic algae [58]. However, previous studies have shown the difference between the $\delta^{13}$C values of algae and its coral hosts is very small when corals derive most of their carbon from their symbionts [59, 60].

**Figure 10** shows the stable carbon isotope distribution of the sampled organisms in Triángulos Reef. $\delta^{13}$C values of the corals ranged between $-19.7$‰ and $-14.8$‰,

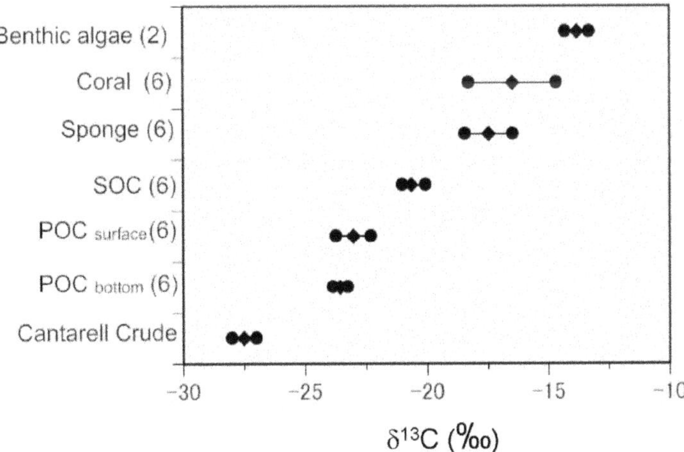


**Figure 10.**
*Stable carbon isotope composition of benthic algae, coral tissue* (Montrastrea cavernosa*) and sponge tissue form Triángulos Reef collected in September 2001. Also shown are the δ*$^{13}$*C values of near-surface particulate and near-bottom organic carbon (POC) and sedimentary organic carbon (SOC) of the region, and of typical Campeche Bank crude [21]. Number of samples are given in parentheses.*

with an average value of $-16.5$‰ (standard deviation of $\pm1.8$‰), which is statistically similar to the average in sponge samples of $-17.4$‰($\pm1.0$) (range of $-19.0$ to $-16.2$‰). In turn, the carbon isotope composition of the two benthic macroalgae samples averaged $-13.8$‰ ($\pm0.5$). The isotopic values of these organisms are significantly different from typical crude oil carbon isotope composition of $-28$ to $-26$‰ of marine siliciclastic and carbonate reservoirs from the Bay of Campeche in the southern Gulf of Mexico [20, 21]. The measured values from our coral samples are similar to those from coral and zooxanthellae ($-20.5$ to $-13.5$‰ in *Stylophora pistillata*; $-19.1$ to $.11.9$‰ in *Favia favus*) from the Red Sea [60], coral tissue from two reefs in the western Pacific Ocean ($-14.6$ to $-12.1$‰ [61]), to $\delta^{13}C$ values between $-16.6$ and $-12.4$‰ in *Montastrea annularis* from Jamaican reefs [59] and lie within the range for several Indo-Pacific and Caribbean reefs [58]. These data further suggest that the principal carbon source is provided by the zooxanthellae, since these corals are relatively enriched in $^{13}C$, in contrast to a $^{13}C$-depleted signature when heterotrophic activity by the coral becomes the principal food source [59, 62].

PAHs (organic compounds with two or more fused aromatic rings) have different sources in the marine environment [63]. Oil-derived PAHs account for about 20% of total hydrocarbons in crude oil and are complex mixtures of two to eight rings although naphthalene and its alkylated homologs are usually present at higher concentrations since, in crude oil, PAH concentrations usually decrease with increasing molecular weight [63]. Other PAH sources to the environment include pyrolysis of organic matter which generates high-molecular-weight PAH [64] and microbial and plant biosyntheses [63].

**Table 5** gives concentrations of individual and total PAH concentrations ($\Sigma$PAH) in coral and sponge samples from this study and for three potential hydrocarbon sources in the region [57]. $\Sigma$PAH averaged 4.5 ppb in coral and 11.7 ppb in sponge (excluding the eastern flank sample with 147.4 ppb). These results suggest a relatively pristine environment, similar to PAH concentrations in reef organisms from Micronesia [65], where $\Sigma_{16}$PAH for sponges and corals varied between 7 and 722 ppb. PAH concentrations of this study are also low than those from the Red Sea coast, where $\Sigma_{16}$PAH in corals was two to three orders of magnitude higher [66]. Compared to coral samples, measurable PAHs in sponge samples reflect their relatively high lipid content and limited PAH-metabolizing capabilities [65]. Except for measurable concentrations of naphthalene and its alkylated homologs, most individual PAH concentrations in coral samples were present below the detection or quantitation limits (**Table 5**). In contrast, three of the six sponge samples showed concentrations around 5 ppb of benzo(*b*)fluoranthene, three had phenanthrene near the detection limit, one sample had 4 ppb fluorine, and the significantly higher PAH concentrations of the sponge collected in the eastern flank (station 6) result from high values of alkylated naphthalenes. Individual PAH distributions suggest an oil-derived origin, where low-molecular-weight PAH and its alkylated homologs predominate [63, 66]. In contrast, the lack of high-molecular-weight PAHs indicates the absence of these compounds from pyrogenic sources.

One month before this study, Cram et al. [57] measured individual PAHs in sediments from Triángulos Reef and nearby Cayo Arcas. These authors also analyzed the PAH composition from Cayo Arcas and Cantarell crude oils and from the ship's fuel oil. They detected measurable PAHs in only 7 of the 71 sediment samples in Cayo Arcas with concentrations between 3 and 28 ppm, and, of the 6 individual PAHs they detected, only pyrene and benzo(*b*)fluoranthene were also found in the sponge samples from the present study (**Table 5**). Tracing the source of oil in the reef's sediments, Cram et al. [57] suggest a pyrolytic origin of PAHs where high-molecular-weight hydrocarbons (4–6 rings) predominate. Another potential source

| Sample | Site | Reef sector | $\delta^{13}$C (‰) | Naphthalene | 2-Methylnaphthalene | 1-Methylnaphthalene | 2,6-Dimethylnaphthalene | Acenaphthene | Fluorene | Dibenzothiophene | Phenanthrene | Anthracene | Fluoranthene | Pyrene | Benzo(a)anthracene | Chrysene | Benzo(b)fluoranthene | Benzo(k)fluoranthene | Benzo(b)pyrene | Benzo(a)pyrene | Perylene | Indene(1,2,3,cd)pyrene | Benzo(g,h,i)perylene | ΣPAH |
|---|---|---|---|---|---|---|---|---|---|---|---|---|---|---|---|---|---|---|---|---|---|---|---|---|
| | | | | | | | | | | | PAH (µg/kg) | | | | | | | | | | | | |
| Coral | 1 | NE | −15.40 | — | 2.4 | 1.4 | — | — | — | — | — | — | — | — | — | — | — | — | — | — | — | — | — | 3.8 |
| | 2 | NE | −19.69 | — | 2.0 | 1.4 | — | — | 1.2 | — | — | — | — | — | — | — | — | — | — | — | — | — | 1.0 | 5.6 |
| | 3 | W | −14.76 | 1.8 | 4.7 | 3.0 | — | — | — | — | — | — | — | — | — | — | — | — | — | — | — | — | — | 9.5 |
| | 4 | SE | −17.09 | — | 1.5 | | — | — | — | — | — | — | — | — | — | — | — | — | — | — | — | — | — | 1.5 |
| | 5 | W | −15.21 | — | 2.2 | 1.3 | — | — | — | — | — | — | — | — | — | — | — | — | — | — | — | — | — | 3.5 |
| | 6 | W | −16.67 | — | 2.0 | 1.1 | — | — | — | — | — | — | — | — | — | — | — | — | — | — | — | — | — | 3.1 |
| Sponge | 0 | | | — | — | — | — | — | — | — | — | — | — | — | — | — | — | — | — | — | — | — | — | — |
| | 1 | NE | −16.96 | — | 2.4 | — | 2.2 | — | — | — | — | — | — | — | — | — | 4.2 | — | — | — | — | — | — | 8.8 |
| | 2 | NE | −17.58 | 2.7 | 4.6 | 2.9 | 3.4 | — | — | — | 2.3 | — | — | — | — | — | 5.8 | — | — | — | — | — | — | 21.7 |
| | 3 | W | −19.03 | — | 3.0 | 2.0 | 2.3 | — | — | — | — | — | — | — | — | — | 5.2 | — | — | — | — | — | — | 12.5 |
| | 4 | SE | −17.98 | — | 1.3 | — | 1.2 | — | — | — | 2.3 | — | 1.7 | 1.3 | — | — | — | — | — | — | — | — | — | 7.8 |
| | 5 | W | −16.24 | — | 2.4 | 1.6 | 2.2 | — | — | — | 1.5 | — | — | — | — | — | — | — | — | — | — | — | — | 7.7 |
| | 6 | W | −16.89 | 29.0 | 58.0 | 33.0 | 19.0 | — | 4.0 | 1.6 | 2.8 | — | — | — | — | — | — | — | — | — | — | — | — | 147.4 |
| Crude oils[57] | CN | | N.A. | 137.0 | N.A. | N.A. | N.A. | 97.0 | 81.0 | N.A. | 93.0 | 121.0 | 17.0 | 50.0 | 50.0 | 17.0 | 16.0 | 14.0 | N.A. | — | N.A. | N.A. | — | 745.0 |
| | CA | | N.A. | 39.3 | N.A. | N.A. | N.A. | 70.0 | 107.8 | N.A. | 45.5 | 273.0 | 128.7 | 264.8 | — | — | — | — | N.A. | — | N.A. | N.A. | — | 1001.6 |

| Sample | Site | Reef sector | δ$^{13}$C (‰) | Naphthalene | 2-Methylnaphthalene | 1-Methylnaphthalene | 2,6-Dimethylnaphthalene | Acenaphthene | Fluorene | Dibenzothiophene | Phenanthrene | Anthracene | Fluoranthene | Pyrene | Benzo(a)anthracene | Chrysene | Benzo(b)fluoranthene | Benzo(k)fluoranthene | Benzo(b)pyrene | Benzo(a)pyrene | Perylene | Indene(1,2,3,cd)pyrene | Benzo(g,h,i)perylene | ΣPAH |
| --- | --- | --- | --- | --- | --- | --- | --- | --- | --- | --- | --- | --- | --- | --- | --- | --- | --- | --- | --- | --- | --- | --- | --- | --- |
| | | | | | | | | | | | PAH (μg/kg) | | | | | | | | | | | | |
| Fuel oil[57] | | | N.A. | 208.6 | N.A. | N.A. | N.A. | 199.1 | 454.4 | N.A. | 111.8 | 127.3 | 77.3 | — | — | — | — | — | N.A. | — | N.A. | N.A. | — | 1772.1 |
| Sfc water | | | | 9.8 | — | — | — | — | — | — | 7.9 | — | 8.0 | 7.9 | — | — | — | — | — | — | — | — | — | — |
| Cayo Arcas sediments[57] | | | N.A. | — | N.A. | N.A. | N.A. | — | — | N.A. | — | P | — | P | P | P | P | — | N.A. | P | N.A. | N.A. | — | — |

*Concentrations in bold are present below the limit of quantitation. Also given are the PAH composition of Cantarell crude oil (CN), Cayo Arcas crude oil (CA), and fuel oil from the sampling ship as reported by Cram et al. [57]. Organisms and sediments were collected in September 2001. Surface water samples were collected in September 2010. N.A., not analyzed; P, present but not quantified.*

**Table 5.**
*Stable carbon isotope values, polycyclic aromatic hydrocarbons (PAHs), and naphthalene-alkylated homolog concentrations (wet weight) in coral (*Montastraea cavernosa*) and sponge samples from Triángulos reef.*

of hydrocarbons appears to be ballast water from the relatively intense ship traffic in the region [57]. Our data, especially the sponge sample from station 6, suggest that a potential source of these compounds in the reef could be the crude oils from the region or discharged ballast waters from oil tankers. The relatively higher solubility of low-molecular-weight PAHs can explain their incorporation into the reef's food web.

In summary, the $\delta^{13}C$ values measured in corals suggest a carbon source from fixation by zooxanthellae. This implies that Triángulos Reef is not under evident stress, because corals expel these algae under unfavorable conditions. In such a case, the isotopic composition of the coral would resemble that of heterotrophic activity, such as feeding from zooplankton. Additionally, most PAHs measured in corals were below the detection limit, while the few individual PAHs detected in sponges and corals (mostly low-molecular-weight PAHs) suggest the presence of oil-derived compounds from either Cantarell oil field or Cayo Arcas marine terminal.

## 4. Conclusions

Stable carbon and nitrogen isotopes from anthropogenically impacted and relatively pristine marine environments in these studies show its usefulness in tracing the sources and flows of these elements in the environment. These studies spanned Gulf of Mexico coastal, shelf, and deep-sea regions, where isotopes were analyzed in surface and subsurface sediments, marine particles, and marine organisms. Ancillary data provided additional information on several biogeochemical issues.

In Bay of Campeche's Cantarell oil field sediments, hydrocarbon seeping was traced at different depths of a sediment core, which showed a distinct $\delta^{13}C$ oil-related signature relative to a nearby reference site. At 20 m core depth, the $^{13}C$-depleted seep sediments were associated with high total hydrocarbon concentrations indicating the vertical migration of oil. In the seep and reference cores, a $\delta^{13}C$ maximum ($-19‰$) is concordant with the onset of the recent interglacial period (10,000 years B.P.).

In another study in this region, carbon and nitrogen isotopes in suspended particles, surface sediments, and macrobenthic organisms from the coastal zone and in the Cantarell oil field were used to trace the sources of organic matter to the benthic community. Results show that marine-derived organic matter is the principal carbon source to Bay of Campeche sediments, with an additional terrestrial contribution in the nearshore region. This implies that, on a shelf-wide scale, coastal discharge, oil extraction, and seeping have no direct effect to the benthic food web.

Carbon and nitrogen stable isotopes measured in surface sediments from the northwest slope of Cuba, along with C:N ratios and $C_{org}$ and TN concentrations, traced the sources of sedimentary organic matter as marine derived. Surface sediment $\delta^{15}N$ values suggest that nitrate diffusing from the thermocline, and not nitrogen fixation, is the principal long-term nitrogen source to primary producers in the upper waters. In addition, a comparison with reported $\delta^{15}N$ and $\delta^{13}C$ values in the region suggests that organic matter flux from the surface ocean to the sediments is fast, most likely as sinking particles.

In Triángulos Reef, the stable carbon isotope composition of corals and sponges suggests that this ecosystem is not under evident stress, because the $\delta^{13}C$ signal would then change to that of a heterotrophic-based carbon source. PAH concentrations suggest a relatively pristine environment, although the individual PAH distribution, with relatively high concentrations of naphthalene, 1-methylnaphthalene, 2-methylnaphthalene, and 2,6-dimethylnaphthalene, indicates some impact from petrogenic sources.

## Acknowledgements

Research summarized in this paper was funded by the Instituto Mexicano del Petróleo through project FIES 98-61-VI (DLV) and by a grant provided by REPSOL-YPF Cuba S.A. (Dr. Luis A. Soto, Instituto de Ciencias del Mar y Limnología, UNAM). I acknowledge Dr. L. Pérez-Cruz (Instituto de Geofísica, UNAM) for providing Bay of Campeche sediment cores and M. Escudero (Instituto Mexicano del Petróleo), Drs. L.A. Cifuentes and J. Kaldy (Department of Oceanography, Texas A&M University), P. Morales (Instituto de Geología, UNAM), and Drs. P.K. Swart and A. Saied (Rosenstiel School of Marine and Atmospheric Sciences, University of Miami) for their isotopic analyses.

## Author details

Diego López-Veneroni[1,2]

1 Instituto Mexicano del Petróleo, México, D.F., México

2 Independent Researcher, México

*Address all correspondence to: dlvmx@yahoo.com

IntechOpen

## References

[1] DeNiro MJ, Epstein S. Influence of diet on the distribution of carbon isotopes in animals. Geochimica et Cosmochimica Acta. 1978;**42**:495-506

[2] DeNiro MJ, Epstein S. Influence of diet on the distribution of nitrogen isotopes in animals. Geochimica et Cosmochimica Acta. 1981;**45**:341-351

[3] Fry B, Sherr EB. $\delta^{13}C$ measurements as indicators of carbon flow in marine and freshwater ecosystems. Contributions in Marine Science. 1984;27:13-47

[4] Vander Zanden MJ, Rasmussen JB. Variation in $\delta^{15}N$ and $\delta^{13}C$ trophic fractionation: Implications for aquatic food web studies. Limnology and Oceanography. 2001;**46**:2061-2066

[5] Minagawa M, Wada W. Stepwise enrichment of $^{15}N$ along food chains: Further evidence and the relation between $\delta$ $^{15}n$ and animal age. Geochimica et Cosmochimica Acta. 1984;**48**:1135-1140

[6] Hoefs J. Stable Isotope Geochemistry. 3rd ed. Berlin: Springer-Verlag; 1987. 241p

[7] Lajtha K, Michener RH, editors. Stable Isotopes in Ecology and Environmental Sciences. Oxford: Blackwell Scientific; 1994. 316p

[8] Fry B. Stable Isotope Ecology. New York: Springer; 2006. DOI: 10.1007/0-387-33745-8. 308p

[9] Phillips DL, Koch PL. Incorporating concentration dependence in stable isotope mixing models. Oecologia. 2002;**130**:114-125

[10] Mateo MA, Serrano O, Serrano L, Michener RH. Effects of sample preparation on stable isotope ratios of carbon and nitrogen in marine invertebrates: Implications for food web studies using stable isotopes. Oecologia. 2008;**157**:105-115

[11] Kennedy P, Kennedy H, Papadimitriou S. The effect of acidification on the determination of organic carbon, total nitrogen and their stable isotopic composition in algae and marine sediments. Rapid Communications in Mass Spectrometry. 2005;**19**:1063-1068

[12] EPA (Environmental Protection Agency). Test Method for Evaluating Total Recoverable Petroleum Hydrocarbons (Method 418.1). Washington, D.C.: U.S. Government Printing Office; 1978

[13] MacDonald IR, Bohrmann G, Escobar E, Abegg F, Blanchon P, Blinova V, et al. Asphalt volcanism and chemosynthetic life in the Campeche Knolls, Gulf of Mexico. Science. 2004;**304**:999-1002

[14] Miranda FP, Quintero-Mármol AM, Campos Pedroso E, Beisl CH, Welgan P, Medrano ML. Analysis of RADARSAT-1 data for offshore monitoring activities in the Cantarell complex, Gulf of Mexico, using the unsupervised Semivariogram textural classifier (USTC). Canadian Journal of Remote Sensing. 2004;**30**:424-436

[15] Quintero-Marmol AM, Pedroso EC, Beisl CH, Caceres RG, de Miranda FP, Bannerman K, et al. Operational applications of RADARSAT-1 for the monitoring of natural oil seeps in the South Gulf of Mexico. In: Geoscience and Remote Sensing Symposium (IGARSS'03) Proceedings, Vol. 4; 2003. pp. 2744-2746

[16] Ortega-Osorio A, Perez-Cruz LL. Geochemistry of marine sediments associated to gas pockets and seeps in the Gulf of Mexico. Paper Presented at

the 2003 AAPG International Conference & Exhibition; Barcelona; 21–24 September, 2003

[17] Hedges JI, van Geen A. A comparison of lignin and stable carbon isotope compositions in quaternary marine sediments. Marine Chemistry. 1982;**11**:43-54

[18] Jasper JP. An organic geochemical approach to problems with glacial-interglacial climatic variability [Dissertation]. Woods Hole: Massachusetts Institute of Technology, Woods Hole Oceanographic Institute; 1998

[19] Meyers PA. Preservation of elemental and isotopic source identification of sedimentary organic matter. Chemical Geology. 1994;**114**: 289-302

[20] Sweeney RE, Haddad RI, Kaplan IR. Tracing the dispersal of the Ixtoc-I Oil using C, H, S, and N stable isotope ratios. In: Proceedings of a Symposium on Preliminary Results from the September 1979 Researcher/Pierce Ixtoc-I Cruise. June 9–10, 1980; Key Biscayne, Florida; 1980. pp. 89-115

[21] Guzmán-Vega A, Mello M. Origin of oil in the Sureste Basin, Mexico. American Association of Petroleum Geologists. 1999;**3**:1068-1095

[22] Gilhooly WP III, Macko SA, Flemings PB. Data report: Isotope compositions of sedimentary organic carbon and total nitrogen from Brazos-Trinity Basin IV (Sites U1322 And U1224), Deepwater Gulf of Mexico. In: Flemings PR, Behrman JH, John CM, and the Expedition 308 Scientists, editors. Proceedings IODP, 308; College Station, Texas; 2018. DOI: 10.2204/iodp. proc.308.208.2008

[23] Newman J, Parker PL, Behrens EW. Organic carbon isotope ratios in quaternary cores from the Gulf of Mexico. Geochimica et Cosmochimica Acta. 1973;**37**:225-238

[24] Clark PU, Dyke AS, Shakun JD, Carlson AE, Clark J, Wohlfath B, et al. The last glacial maximum. Science. 2009;**325**:710-714

[25] Kohl B, Williams DF, Ledbetter MT, Constans RE, King JW, Heuser LE, et al. Summary of chronostratigraphic studies, Deep Sea Drilling Project, Leg 96'. In: Bouma AH, Coleman JM, Meyer AW, editors. Initial Reports of the Deep Sea Drilling Project, 96; Washington; 1986. pp. 589-600

[26] Yeager KM, Santschi PH, Rowe GT. Sediment accumulation and radionuclide inventories ($^{239,240}$Pu, $^{210}$PB and $^{234}$Th) in the Northern Gulf of Mexico, as influenced by organic matter and macrofaunal density. Marine Chemistry. 2004;**91**:1-14

[27] Poore RZ, Dowsett HJ, Verardo S, Quinn TM. Millennial- to century-scale variability in Gulf of Mexico Holocene climate records. Paleoceanography. 2003;**18**(2):26-1-26-13. DOI: 10.1029/ 2002PA000868

[28] Joyce JE, Tjalsma LRC, Prutzman JM. High-resolution Planktic Stable Isotope Record and Spectral Analysis for the Last 5.35 M.Y.: Ocean Drilling Program Site 625 Northeast Gulf of Mexico. Paleoceanography. 1993;**5**:507-529

[29] Signoret M, Monreal-Gómez MA, Aldeco J, Salas-de-León DA. Hydrography, oxygen saturation, suspended particulate matter, and chlorophyll-a fluorescence in an oceanic region under freshwater influence. Estuarine, Coastal and Shelf Science. 2006;**69**:153-164

[30] Zavala-Hidalgo J, Gallegos-García A, Martínez-López B, Morey SL, O'Brien JJ. Seasonal upwelling on the western and southern shelves of the

Gulf of Mexico. Ocean Dynamics. 2006;
**56**:333-338

[31] Hidalgo-González RM, Alvarez-Borrego S. Water column structure and phytoplankton biomass profiles in the Gulf of Mexico. Ciencias Marinas. 2008; **34**:197-212

[32] Salas-de-León DA, Monreal-Gómez MA, Signoret M, Aldeco J. Anticyclonic-cyclonic eddies and their impact on near-surface chlorophyll stocks and oxygen supersaturation over the Campeche canyon, Gulf of Mexico. Journal of Geophysical Research. 2004; **109**(C5):1-10. DOI: 10.1029/2002JC001614

[33] Ortiz-Zamora G, Huerta-Díaz M, Salas-de-León DA, Monreal-Gómez MA. Degrees of pyritization in the Gulf of Mexico in sediments influenced by the Coatzacoalcos and the Grijalva-Usumacinta Rivers. Ciencias Marinas. 2006;**28**:369-379

[34] Hernandez-Arana HA, Attrill MJ, Hartley R, Gold-Bouchot G. Transitional carbonate-terrigenous shelf sub-environments inferred from textural characteristics of surficial sediments in the southern Gulf of Mexico. Continental Shelf Research. 2005;**25**: 1836-1852

[35] Sánchez-García S. Determinaciones isotópicas de carbono y nitrógeno en sedimentos y biota asociados a emanaciones naturales de hidrocarburos fósiles en el Banco de Campeche, México [Carbon and nitrogen isotopic determinations in sediments and biota associated to natural fossil hydrocarbon seeps in Bank of Campeche, Mexico] [MSc tesis]. Mexico City: Instituto de Ciencias del Mar y Limnología, Universidad Nacional Autónoma de México, México; 2003

[36] Wade T, Kennicutt MC II, Brooks JM. Gulf of Mexico hydrocarbon seep communities: Part III. Aromatic hydrocarbon concentrations in organisms, sediments and water. Marine Environmental Research. 1989; **27**:19-30

[37] Eadie BJ, Jeffrey LM. δ$^{13}$C analyses of oceanic particulate organic matter. Marine Chemistry. 1973;**1**:199-209

[38] Benner R, Opsahl S. Molecular indicators of the sources and transformations of dissolved organic matter in the Mississippi River plume. Organic Geochemistry. 2001;**31**:597-611

[39] López-Veneroni D. The dynamics of dissolved and particulate nitrogen in the Northwest Gulf of Mexico [Dissertation]. College Station: Texas A&M University; 1998

[40] Botello AV, Villanueva S, Mendelewicz M. Programa de Vigilancia de los hidrocarburos en sedimentos del Golfo de México y Caribe Mexicano 1978–1984 [Hydrocarbon Survellance Program in Gulf of Mexico and Mexican Caribbean Sediments 1978–1984]. Caribbean Journal of Science. 1987;**23**: 29-39

[41] Soto LA, Escobar E. Coupling mechanisms related to benthic production in the SW Gulf of Mexico. In: Eleftheriou A, Ansell AD, Smith CJ, editors. Biology and Ecology of Shallow Coastal Waters. Fredensborg: Olsen & Olsen; 1995. pp. 233-242

[42] Michener RH, Kaufman L. Stable isotope ratios as tracers in marine food webs: An update, In: Coupling mechanisms related to benthic production in the SW Gulf of Mexico. In: Lajtha K, Michener RH, editors. Stable Isotopes in Ecology and Environmental Sciences. Oxford: Blackwell Scientific; 1994. pp. 238-278

[43] Salcedo DL, Soto LA, Paduan JB. Trophic structure of the macrofauna associated to deep-vents of the southern Gulf of California: Pescadero Basin and

Pescadero Transform Fault. PLoS One. 2019;**14**:e0224698

[44] Okolodkov YB. A review of Russian plankton research in the Gulf of Mexico and the Caribbean Sea in the 1960-1980s. Hydrobiologia. 2003;**13**: 207-221

[45] Macko SA, Entzeroth L, Parker PL. Regional differences in the nitrogen and carbon isotopes on the continental shelf of the Gulf of Mexico. Die Naturwissenschaften. 1984;**71**:374-375

[46] Ellwood BB, Balsamand WL, Roberts HH. Gulf of Mexico sediment sources and sediment transport trends from magnetic susceptibility measurements of surface samples. Marine Geology. 2006;**230**:237-248

[47] Magnier C, Moretti I, Lopez JO, Gaumet F, Lopez JG, Letouzey J. Geochemical characterization of source rocks, crude oils and gases of Northwest Cuba. Marine and Petroleum Geology. 2004;**21**:195-214

[48] Soto LA, de la Lanza G, López-Veneroni D. Depth profiles of stable nitrogen and carbon isotopes and C:N ratios in surficial sediments from the NW insular slope of Cuba. Earth Ocean and Space: Transactions American Geophysical Union. 2007; **88**:23. Jt. Assem, Suppl., Abstract OS53B-02

[49] Soto LA, López-Veneroni D, López-Canovas D, Ruíz-Vázquez R, de la Lanza G. Surface sediment survey of the seabed on the Northwestern Slope of Cuba, Southern Straits of Florida. Interciencia. 2012;**37**:812-819

[50] Goñi MA, Ruttenberg KC, Eglington TI. A reassessment of the sources and importance of land-derived organic matter in surface sediments from the Gulf of Mexico. Geochimica et Cosmochimica Acta. 1998;**62**:3055-3075

[51] Abascal AJ, Sheinbaum J, Candela J, Ochoa J, Badan A. Analysis of flow variability in the Yucatan Channel. Journal of Geophysical Research. 2003; **108**(C12):11-1-11-18. DOI: 10.1029/ 2003JC001922

[52] Mullholland MR, Bernhardt PW, Heil CA, Bronk DA, O'Neil JM. Nitrogen fixation and release of fixed nitrogen by *Trichodesmium* spp. in the Gulf of Mexico. Limnology and Oceanography. 2006;**51**:1762-1776

[53] Sigman DM, Casciotti KL. Nitrogen isotopes in the ocean. In: Steele JH, Turekian KK, Thorpe SA, editors. Encyclopedia of Ocean Sciences. San Diego: Academic; 2001. pp. 1884-1894

[54] Altabet MA. Variations in nitrogen isotopic composition between sinking and suspended particles: Implications for nitrogen cycling and particle transformation in the open ocean. Deep Sea Research. 1998;**35**:535-554

[55] Lane A, Harrison PL. Effects of oil contaminants on survivorship of larvae of the scleractinian reef corals *Acropora tenuis*, *Goniastrea aspera* and *Platygyra sinensis* from the Great Barrier Reef. In: Proceedings 9th International Coral Reef Symposium; Bali, Indonesia; 23–27 October, 2000

[56] Jordán-Dahlgren E. Coral reefs of the Gulf of Mexico: Characterization and diagnosis. In: Withers K, Nipper M, editors. Environmental Analysis of the Gulf of Mexico. Corpus Christi: Harte Institute for Gulf of Mexico Studies; 2007. pp. 340-350

[57] Cram S, Ponce de León CA, Fernández P, Sommer I, Rivas H, Morales LM. Assessment of trace elements and organic pollutants form a marine oil complex into the coral reef system of Cayo Arcas, Mexico. Environmental Monitoring and Assessment. 2006;**121**:127-419

[58] Heikoop JM, Dunn JJ, Risk MJ, Tomascik T, Schwarcz HP, Sandeman IM, et al. $\delta^{15}N$ and $\delta^{13}C$ of coral tissue show significant inter-reef variation. Coral Reefs. 2000;**19**:189-193

[59] Land LS, Lang JC, Smith BN. Preliminary observations on the carbon isotopic composition of some reef coral tissues and symbiotic zooxanthellae. Limnology and Oceanography. 1975;**20**: 283-287

[60] Alamaru A, Loya Y, Brokovich E, Yam R, Shemesh A. Carbon and nitrogen utilization in two species of Red Sea corals along a depth gradient: Insights from stable isotope analysis of total organic material and lipids. Geochimica et Cosmochimica Acta. 2009;**73**:5333-5342

[61] Yamamuro M, Kayannke H, Minagawa M. Carbon and nitrogen stable isotopes of primary producers in coral reef ecosystems. Limnology and Oceanography. 1995;**40**:617-621

[62] Swart PK, Saied A, Lamb K. Temporal and spatial variation in the $\delta^{15}N$ and $\delta^{13}C$ of coral tissue and zooxanthellae in Montastraea faveolata collected from the Florida reef tract. Limnology and Oceanography. 2005;**50**: 1049-1058

[63] Neff JM. Polycyclic Aromatic Hydrocarbons in the Aquatic Environment—Sources, Fates, and Biological Effects. London: Applied Sciences Publishers; 1979

[64] Poulsen A, Burns K, Lough J, Brinkman D, Delean S. Trace analysis of hydrocarbons in coral cores from Saudi Arabia. Organic Geochemistry. 2006;**37**: 1913-1930

[65] Denton GRW, Concepcion LP, Wood HR, Morrison RJ. Polycyclic aromatic hydrocarbons (PAHs) in small island coastal environments: A case study from harbours in Guam,

Micronesia. Marine Pollution Bulletin. 2006;**52**:1090-1117

[66] El-Sikaily A, Khaled A, El Nemr A, Said TO, Abd-Alla AMA. Polycyclic aromatic hydrocarbons and aliphatics in the coral reef skeleton of the Egyptian Red Sea coast. Bulletin of Environmental Contamination and Toxicology. 2003;**71**:1252-1259

# Benthic Macroinvertebrate Communities as Indicators of the Environmental Health of the Cunas River in the High Andes, Peru

*María Custodio, Richard Peñaloza and Heidi De La Cruz*

## Abstract

The Cunas River is a valuable natural freshwater heritage in the central region of Peru, where diverse economic activities depend on the quantity and quality of its waters. The environmental health of the Cunas River was assessed through indicators of the diversity of benthic macroinvertebrate communities and multivariate statistical methods. Water and sediment samples were collected in sectors of three populated centers during 2017. Indicators of water quality and diversity of benthic macroinvertebrates were determined. The results reveal that most of the water quality indicators are in the range of the water quality standards of rivers in Peru. Twenty-six families of benthic macroinvertebrates were identified. The principal component analysis (PCA) of the water quality indicators through the first two components explained 79.59% of the total variance. Cluster analysis in relation to the relative abundance of benthic macroinvertebrates grouped the sampling sites into groups with similar characteristics. Principal coordinate analysis (PCO) analysis of benthic macroinvertebrate communities showed a clear separation of sites. The percentage similarity (SIMPER) analysis at the family level showed the percentage of contribution of species to the benthic fauna community. The canonical correspondence analysis (CCA) identified water quality variables that influence the distribution of benthic macroinvertebrate communities. Therefore, the information obtained will be useful for the management of similar rivers.

**Keywords:** benthic macroinvertebrates, diversity, water quality, river, multivariate analysis

## 1. Introduction

Benthic macroinvertebrates are found in all types of aquatic environments, where they are important indicators of the health of these ecosystems [1]. They inhabit the river bed (among stones, submerged aquatic plants, etc.) either during their entire biological cycle as mollusks or part of it as many insects, in which the adult phase is terrestrial and the larval aquatic phases. Benthic macroinvertebrates have a high variety of morphological and behavioral adaptations in order to take advantage of the different trophic resources offered by a fluvial ecosystem [2, 3].

The composition and structure of benthic macroinvertebrate communities are affected not only by anthropogenic stressors but also by natural factors [4]. In lotic systems, the composition and structure of these communities are controlled by biotic factors (biological interactions: predation, parasitism, competition, etc.) and abiotic factors (water velocity, temperature, discharges, among others) [5, 6]. However, the altitudinal gradient is also considered a determining factor in the distribution of these communities [7]. Although some authors point out that both temperature and oxygen partial pressure are key factors in the distribution of benthic macroinvertebrate communities in river systems [8]. Others report that the integrity of these communities depends on the structural integrity of the current and the processes associated with the physical habitat [9].

Knowledge of benthic fauna in high Andean fluvial ecosystems in the central region of Peru is still scarce considering the large number of continental aquatic ecosystems that exist. The best studied benthic macroinvertebrate communities are located in the high Andean regions of the north of the country compared to the studies carried out in the high Andean regions of central Peru. However, the studies focus on the use of benthic fauna as bioindicators of water quality in monitoring and evaluation programs, since through the analysis of the composition and structure of benthic macroinvertebrate communities, it is possible to determine the degree of disturbance that a body of water has been experiencing.

This study focuses on the Cunas River, one of the most important rivers in the Mantaro River Basin in the Central Andes of Peru. It is 101.1 km long and is located in the provinces of Chupaca, Concepción, Huancayo, and Jauja in the Junín region. In the sub-basin of Cunas River, several economic activities are developed, such as livestock, agriculture (Andean tubers, corn, and vegetables, among others), aquaculture, electricity generation, and the extraction of aggregates (sand and stone). Most of these activities take place without environmental criteria and are exerting strong pressure on the aquatic systems, affecting water quality and the composition of the biological communities. In this sense and considering the high uncertainty about the current health of this aquatic ecosystem, the objective of the study was to evaluate the environmental health of the river Cunas through indicators of water quality and diversity of benthic macroinvertebrates and multivariate statistical methods in precipitation and drought seasons.

## 2. Materials and methods

### 2.1 Description of study area

The Cunas River is located in the central highlands of Peru, in the Mantaro River watershed. It has a length of 101.1 km and is born in the Runapa-Huañunán lagoon at 4535 masl, near the watershed of the Cañete river (western chain). It is located in the provinces of Chupaca, Concepción, Huancayo, and Jauja in the Junín region. Its main channel describes the form of the letter S, with the direction of route west-east. The flow of the river varies according to the time of year. During the rainy season, the flow reaches 152.95 m$^3$/s, and during the dry season, it reaches 2.57 m$^3$/s [10]. Three sampling sectors were defined in the River Cunas, according to their representativeness of the area in terms of the influence of anthropic activity. Sector 1 was located in the town of San Blas, Concepción province, at 3440 masl (18 L 455952E 8670268S), sector 2 in Huarisca at 3315 masl (18 L 471711E 8667535S), and sector 3 in La Perla at 3229 masl (18 L 470205E 8667164S), the latter two in Chupaca

Province (**Figure 1**). In the Cunas River basin, various economic activities are developed, such as agriculture, livestock, aquaculture, tourism, and nonmetallic mining. These activities are exerting strong pressure on the aquatic environment, as there are few efforts to protect this resource.

## 2.2 Sample collection and analysis

Water sampling was carried out in three sectors of the San Blas, Huarisca, and La Perla population centers during 2017. In each sector, ten sampling sites were defined, and in each one of them, pH, conductivity, turbidity, dissolved oxygen (DO), temperature, and dissolved total solids (DTS) were determined in situ using the multiparameter probes Hanna Instruments (HI 991301 Microprocessor pH/temperature, HI 9835 Microprocessor Conductivity/DTS, and HI 9146 Microprocessor dissolved oxygen). Previously, the equipment was calibrated in the respective sampling site. Also, 1 L of water from a depth of 20 cm from the river surface, in the opposite direction to the current flow, was collected from each sampling site for bacteriological analysis of nitrates, phosphates, and $BOD_5$, in containers previously sterilized and treated with a 1:1 solution of hydrochloric acid and rinsed with distilled water. These parameters were measured according to standard methods [11].

The samples were collected using a Surber net with a square frame of $30 \times 30$ cm side (0.09 m$^2$ area) and a 250-μm mesh aperture. Sampling was performed by placing the mesh against the current and removing the substrate upstream of the sleeve [12]. The samples were preserved in 70% alcohol and transferred to the laboratory for identification. Taxonomic identification of benthic

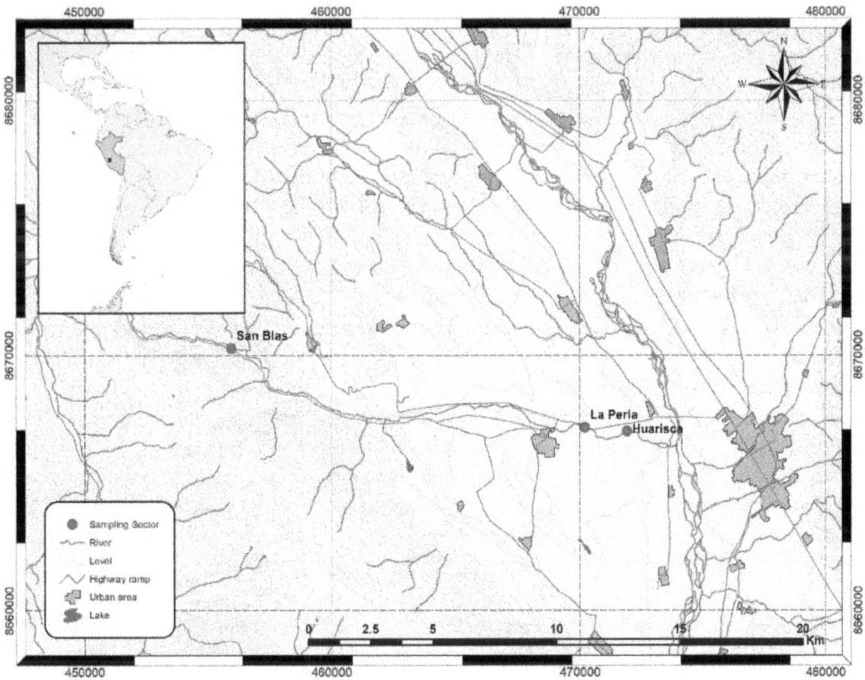


**Figure 1.**
*Map of the location of the sampling sectors in the river Cunas.*

macroinvertebrates was performed at the family level through a trinocular stereomicroscope [13].

## 2.3 Statistical analysis

The analysis of water quality variables was determined by normalized principal component analysis (PCA) in order to generate two-dimensional management maps [14] and search for best-fit lines according to the calculated PCs, successively maximizing the variance of the projected sampling points along each axis. Statistical significance was performed by analyzing the multivariate variance using PERMANOVA permutations [15].

In the analysis of benthic macroinvertebrate communities, a hierarchical and agglomerative classification (cluster analysis) was performed, generating a similarity matrix with Bray-Curtis indices based on an abundance matrix of species transformed by square root in order to produce a dendrogram [16], while a principal coordinate analysis (PCO) was performed to produce a management graph [15]. It was characterized by species richness (S), individual density (N), Shannon diversity index (H′), and Simpson index (1-λ). The main indicator species and the associated percentage indication were determined for each significant set of species, using the percentage similarity (SIMPER) analysis [17]. Canonical correspondence analysis (CCA) was used to evaluate the relationship between water quality variables and macroinvertebrate composition.

## 3. Results

### 3.1 Water quality based on physical, chemical, and bacteriological indicators

The pH of the water presented means and standard deviation that oscillated from 6.99 ± 0.03 in the sector of the Huarisca populated center in the rainy season to 7.59 ± 0.04 in San Blas in the dry season. The highest electrical conductivity (EC) was recorded in the San Blas sector with an average of 567.70 μS/cm. The biochemical oxygen demand (BOD) registered in the La Perla sector surpassed the water quality standards of Peru, destined for human consumption and conservation of the aquatic environment (5 and 10 mg/L, respectively). The water bodies of this same sector presented the lowest concentrations of dissolved oxygen, at both times, as well as the highest temperature. The highest average of dissolved total solids was recorded in the La Perla sector with 377.40 mg/L. The average of phosphates as opposed to nitrates exceeded the water quality standards of the Peruvian Ministry of the Environment for the two types of use considered in this study in the Huarisca and La Perla sectors (**Table 1**).

**Figure 2** shows the result of PCA of the water quality indicators and the sampling sectors, according to towns. The first two components explain 79.59% of the total variance. The first principal component explained 50.76% of the variance and correlated significantly with BOD, DTS, phosphates, nitrates, and thermotolerant coliforms. The second component explained 19.83% of the variance and correlated with pH and EC. Also, the distributions of the groups in the perceptual map show a clear differentiation of the sampling sectors with respect to the main variables. The sectors evaluated in the dry season present higher values of the variables with greater weight in the first two components than their peers in the rainy season, such as the La Perla sector that shows high values, mainly in the PC1 variables. The anthropogenic pressure experienced by the water bodies in the sampling sectors of the middle and lower part of the river Cunas would determine the increase of these variables. In addition, the PERMANOVA results at a significance level of 0.01 show that the observations differ significantly, according to the sampling sector and climatic season factors. That is to say, there is enough statistical evidence to affirm that the sectors have different ranges in relation to the water quality indicators.

| Indicator climate season | San Blas | | Huarisca | | La Perla | |
| --- | --- | --- | --- | --- | --- | --- |
| | Dry | Rainy | Dry | Rainy | Dry | Rainy |
| pH | 7.59 ± 0.04 | 7.06 ± 0.03 | 7.21 ± 0.06 | 6.99 ± 0.03 | 7.41 ± 0.05 | 7.12 ± 0.03 |
| EC (µS/cm) | 567.7 ± 42.23 | 462.6 ± 62.72 | 526.0 ± 36.23 | 482.8 ± 76.24 | 534.2 ± 20.48 | 465.7 ± 49.47 |
| $BOD_5$ (mg/L) | 5.47 ± 0.50 | 4.75 ± 0.87 | 9.24 ± 0.85 | 7.78 ± 0.73 | 11.92 ± 1.04 | 10.09 ± 1.26 |
| Turbidity (NTU) | 2.03 ± 0.40 | 4.85 ± 1.17 | 3.69 ± 1.18 | 14.96 ± 1.74 | 6.10 ± 0.75 | 24.21 ± 2.42 |
| DO (mg/L) | 6.95 ± 1.13 | 6.60 ± 0.78 | 7.13 ± 1.03 | 6.56 ± 0.84 | 5.94 ± 0.91 | 4.19 ± 0.53 |
| Temperature (°C) | 16.62 ± 0.46 | 15.11 ± 1.24 | 18.58 ± 0.92 | 14.62 ± 1.12 | 19.61 ± 0.94 | 15.88 ± 0.82 |
| DTS (mg/L) | 166.2 ± 4.66 | 120.7 ± 8.85 | 299.9 ± 6.42 | 284.1 ± 7.20 | 377.4 ± 8.21 | 352.7 ± 11.99 |
| Phosphates (mg/L) | 0.01 ± 0.00 | 0.021 ± 0.01 | 0.117 ± 0.02 | 0.072 ± 0.02 | 0.242 ± 0.01 | 0.118 ± 0.02 |
| Nitrates (mg/L) | 0.03 ± 0.01 | 0.02 ± 0.01 | 0.05 ± 0.01 | 0.04 ± 0.01 | 0.23 ± 0.03 | 0.06 ± 0.01 |

**Table 1.**
*Mean and standard deviation of water quality indicators of the river Cunas, according to population center and climate season.*

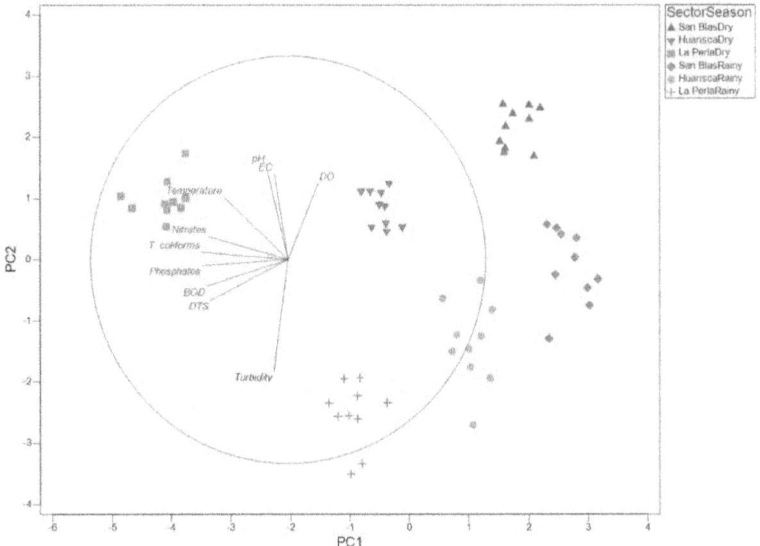


**Figure 2.**
*Perceptual map of principal component analysis (PCA) based on the water quality indicators of the river Cunas.*

## 3.2 Spatial and temporal variation of benthic macroinvertebrate communities

A total of 26 families of benthic macroinvertebrates were found during the two sampling seasons in San Blas, Huarisca, and La Perla sectors. The Diptera order was the most representative in abundance and richness. PCO of the composition of the benthic macroinvertebrate communities showed a clear separation of sites, mainly due to the effect of the season factor (**Figure 3**). The first management axis shows the significant separation of sectors in relation to families and number of individuals. It also shows that the groups are clearly delimited, which explains the percentage of total variation of the first two coordinates (57.02%), separating the sectors into two main groups characterized by the climatic season factor. The analysis shows that there is a high similarity in the community of benthic macroinvertebrates of the La Perla sector in the dry season for axis 1 with values ranging from 20.033 to 29.99 according to the similarity range of Bray-Curtis, making this assemblage of samples grouped by family's similarity significantly different from the others. However, this does not demerit that the other groups keep specific characteristics that make each sector keep particular characteristics that need to be studied individually. The results also reveal that the Huarisca sector in the rainy season is the most depressed in values of the number of families and individuals. The cluster analysis of benthic macroinvertebrate communities at the family level by Bray-Curtis distance range shows similar and significant associations. This is supported by the analysis of main coordinates (**Figure 3**), in which two differentiated groups are found, one with 40% similarity, explained by the climatic season factor, and the other with 60% similarity of the groups, as observed in the sector of San Blas for the rainy and low seasons, which indicates uniformity in the distribution of species (**Figure 4**).

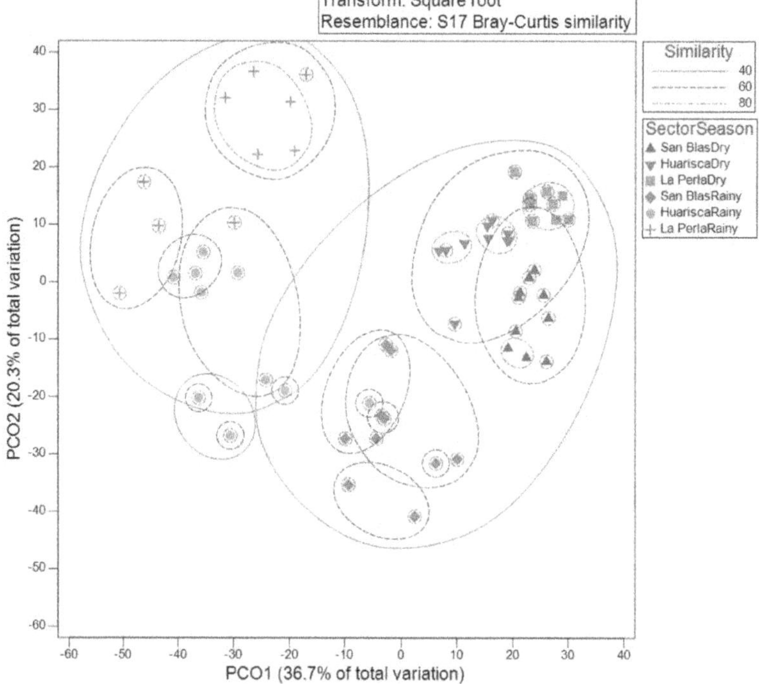


**Figure 3.**
*Principal coordinates analysis (PCO) based on the number of families and abundances of benthic macroinvertebrates of the three sampling sectors, according to sampling season.*

The nonmetric multidimensional scaling analysis shows an average stress level value of 0.16, which according to the range given by Kruskal indicates an acceptable interpretation in the perceptual map. In addition, the high values in nitrates, phosphates, temperature, and thermotolerant coliforms would be conditioning the presence of a greater number of individuals, as can be observed in the La Perla sector during the dry season (**Figure 5**).

The results of the composition of the benthic macroinvertebrate community in the sampling sectors obtained by SIMPER analysis at family level showed that the

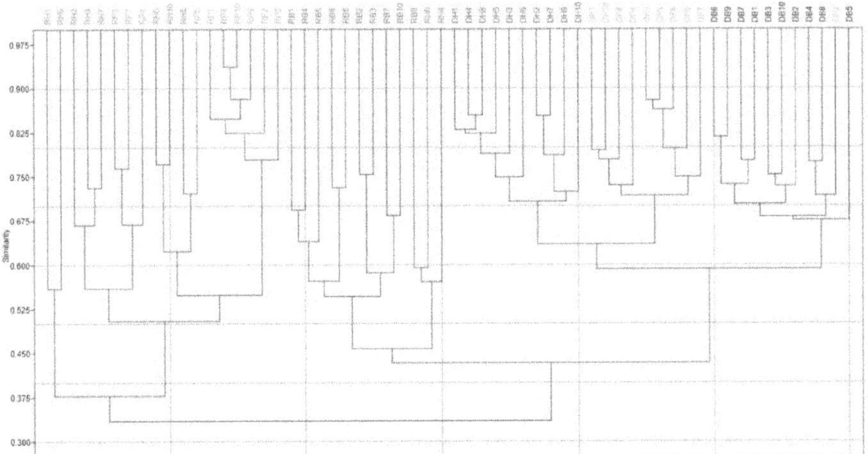

**Figure 4.**
*Dendrogram based on the distances of Bray-Curtis from the benthic macroinvertebrate community of the Cunas River, according to sector and sampling season.*

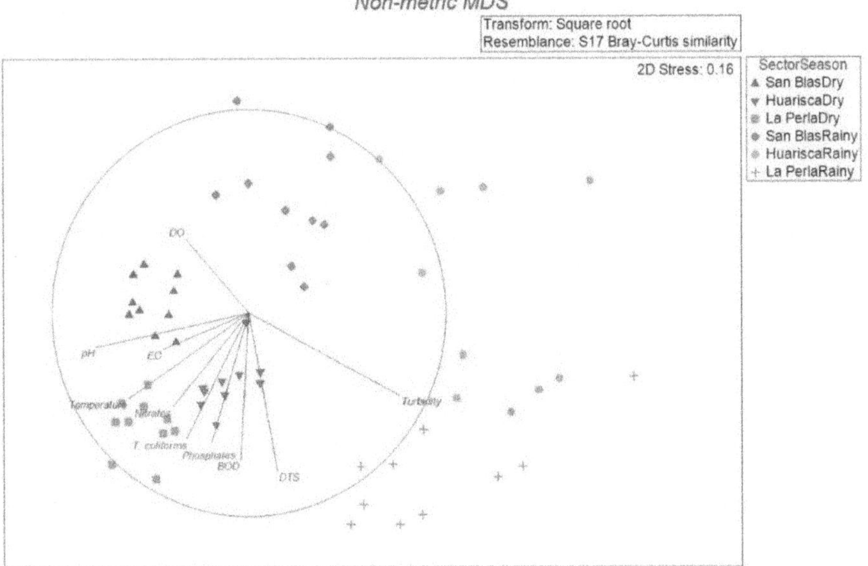


**Figure 5.**
*Analysis of nonmetric multidimensional scaling based on the richness and abundance of benthic macroinvertebrates of the three sampling sectors, according to sampling season.*

| Sampling sector | Taxa | Contribution% | | Diversity indicators | | | |
|---|---|---|---|---|---|---|---|
| | | Dry | Rainy | S | N | H′ | 1-λ |
| San Blas | Baetidae | 28.60 | 40.50 | 26 | 2741 | 1.83 | 0.26 |
| | Chironomidae | 24.81 | 29.08 | | | | |
| | Simuliidae | 10.73 | 11.45 | | | | |
| | Elmidae | 9.81 | | | | | |
| Huarisca | Chironomidae | 44.46 | 73.11 | 22 | 2218 | 1.31 | 0.48 |
| | Simuliidae | 16.45 | | | | | |
| | Baetidae | 12.60 | | | | | |
| La Perla | Chironomidae | 52.29 | 98.42 | 14 | 5394 | 0.77 | 0.74 |
| | Baetidae | 21.60 | | | | | |

*S, number of families; N, number of individuals; H′, Shannon-Wiener index; 1-λ, Simpson index.*

**Table 2.**
*Percentage of the contribution of benthic macroinvertebrate families obtained through SIMPER analysis and mean of diversity indicators.*

highest percentages of contribution in the San Blas sector were made by individuals from Baetidae (40.50%), followed by Chironomidae (29.08%) and Elmidae (11.45%), contributing 81.02% of the total taxa in the rainy season. With respect to diversity indicators, the San Blas sector presented the highest richness and diversity. The results also show that the most dominant family in the Huarisca and La Perla sectors was Chironomidae, with high contribution percentages in both sampling periods (**Table 2**). However, during the rainy season, the Chironomidae reached the highest percentage of contribution in the composition of the benthic macroinvertebrate communities of the Huarisca and La Perla sectors, with 73.11 and 98.42% of the total taxa.

### 3.3 Relationship of water quality and variation patterns of benthic macroinvertebrate communities

The CCA of the water quality and diversity variables of benthic macroinvertebrates shows the new canonical axes extracted and their relationship with the significant water quality variables. In the San Blas sector for both climatic seasons, the largest number of species fits the first axis and has a greater affinity for high EC, DO, and pH values, while the Huarisca and La Perla sectors for the rainy season tend to have less diversity (**Figure 6**).

The results of the matrix similarity test of the water quality variables and benthic macroinvertebrates showed a Spearman correlation coefficient of 66.4%. The best analysis of BIOENV, taking into account the 10 variables under study, shows that turbidity is the variable that has the highest correlation value with the distribution of biological data, with a 60% value in the Spearman range.

The result of the distance-based redundancy analysis (dbRDA) of the variables of water quality and relative abundance of benthic macroinvertebrates is presented in **Figure 7**. The first axis of the redundancy analysis explains 33.0% of the total variance and the second axis 15.7%. The first axis of the dbRDA coordinate shows a higher load for turbidity and pH. It also shows that the values of nitrates, thermotolerant coliforms, and pH are higher in the La Perla and San Blas sectors in the dry season.

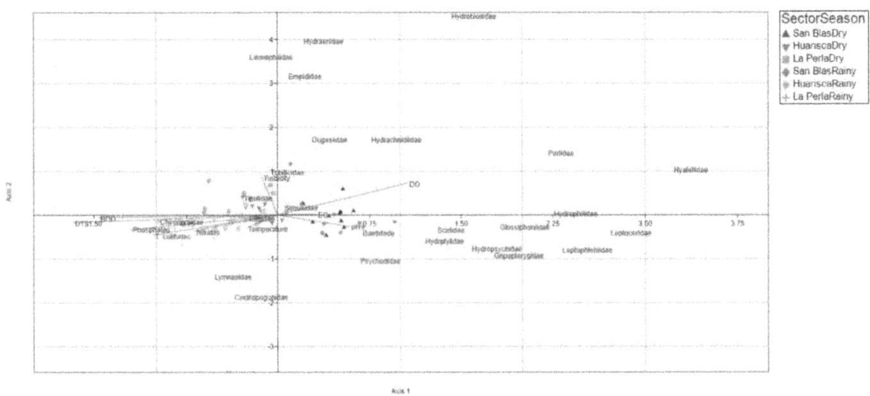

**Figure 6.**
*Analysis of the canonical correspondence of the variables of water quality and diversity of benthic macroinvertebrates of the river Cunas.*

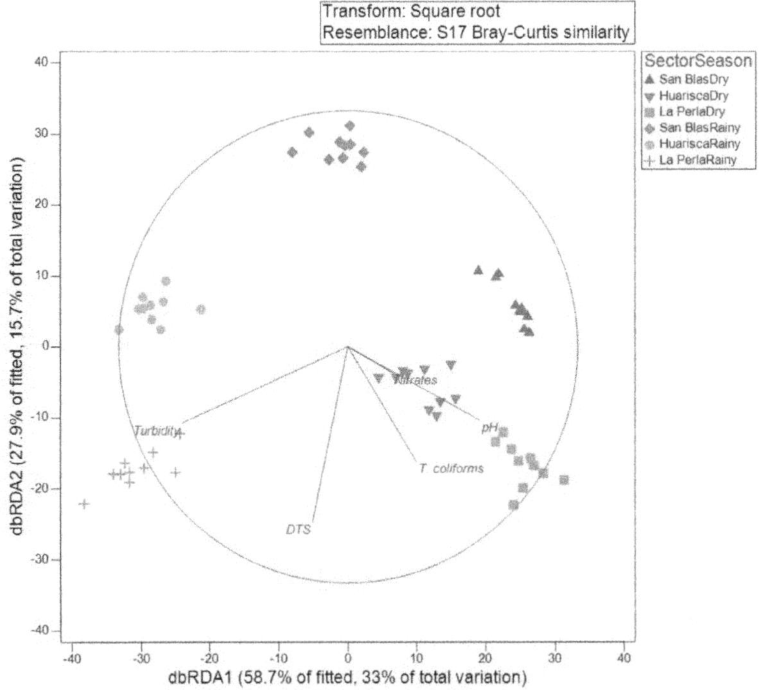


**Figure 7.**
*Perceptual map of the distance-based redundancy analysis (dbRDA) of better physicochemical predictors on the composition of benthic macroinvertebrate communities in the river Cunas.*

## 4. Discussion

### 4.1 Water quality based on physical, chemical, and bacteriological indicators

The results obtained from the evaluation of water quality in the sampling sectors of the river Cunas reveal a progressive deterioration downstream from the headwaters of the basin. This behavior is due to the increase of anthropogenic activities due to the accelerated population growth and migration to urban areas in the region. The higher values of conductivity, BOD recorded in the La Perla sector, are due to the high loads of organic matter in untreated wastewater from different sources [18]. In this sector, BOD values exceeded ~~by far~~ the quality standards of water destined for the conservation of aquatic life, the production of drinking water, and other uses of Peruvian norms [19], as well as the ranks established by the World Health Organization [20] and the Canadian Council of Ministers of the Environment [21].

The results obtained through the PCA reveal that the Cunas River has been experiencing a process of worsening water quality. This is due to the strong anthropogenic activities such as aquaculture in the middle part of the river (San Blas), nonmetallic mining throughout the river course (extraction of aggregates), and discharge of wastewater from nearby urban settlements. The La Perla sector has a poor water quality with respect to BOD and DO. The low concentration of BOD is due to the consumption of this gas in the biodegradation processes, as shown by the high concentrations of BOD registered in this sector. These results are supported by Ayandiran et al. [22], who state that the low oxygen concentration is related to the strong activity of microorganisms that require large amounts of oxygen to metabolize and degrade organic matter. However, another determining factor of oxygen dissolution is temperature, since it determines the tendency of its physical properties, as well as the wealth and distribution of biological communities [23].

Nutrients such as phosphorus in aquatic environments limit the growth of algae and plants, so their determination allows detection of eutrophication problems [24]. The average total phosphorus values obtained in the Huarisca and La Perla sectors exceed the environmental quality standards for the conservation of the aquatic environment (0.035 mg/L). This increase would be related to wastewater discharges, the contribution of detergents, and the drainage of fertilized agricultural soils [25], since the marginal strip of a large part of the river is cultivated areas. In the case of the La Perla sector, the results obtained allow us to classify this body of water in a hypertrophic state with a great algal bloom. In addition, these high concentrations of phosphorus reveal the pollution events through which this sector of the river crosses due to the strong pressure exerted by anthropogenic activities, among them, livestock activities, since cattle feces are a potential source of phosphorus. The mean nitrate concentration values did not exceed the environmental quality standards. In addition, the interaction between phosphorus and iron, at low DO concentrations, results in the release of phosphorus attached to the water column, increasing its concentrations [26].

## 4.2 Spatial and temporal distribution of benthic macroinvertebrate communities

The most abundant benthic macroinvertebrates corresponded to individuals of the class Insecta, order Diptera. The results also reveal significant differences between the macroinvertebrate communities of the evaluated sectors, being the Chironomidae family the most representative with a wide range of distribution [27], in the three altitudinal floors where the sampling sectors were established. As for the contribution of benthic macroinvertebrate families to community composition, the Chironomidae family was consolidated as one of the most important families in the three sampling sectors. The results also reveal that benthic macroinvertebrate communities are dominated by Chironomidae, Simuliidae, and Baetidae families. The abundance of these families confirms the average level of oxygenation of the water masses in the sectors of the river Cunas studied. However, the abundance of the Baetidae family in the San Blas sector indicates that the water masses are oligotrophic, as these organisms usually live at this type of trophic level.

However, the decline of the Baetidae occurs downstream due to low oxygenation levels. These results coincide with those recorded in other studies in aquatic environments with low oxygen levels, where the dominance is of the family Chironomidae [28]. In addition, the dominance of this family in aquatic environments is related to the decrease in water quality, food quality, and interference with breathing mechanisms [29].

This study demonstrates the significant correlation between water quality and benthic macroinvertebrate diversity indicators. These results coincide with those of Verdonschot et al. [30] and Mykrä et al. [31], who report that in temperate climate zones, seasonality plays a vital role in the structure of macroinvertebrate communities. However, the results obtained through the analysis with multivariate methods reveal that the high values in nitrates, phosphates, temperature, and thermotolerant coliforms would be conditioning the presence of a greater number of individuals of the family Chironomidae, resilient to organic pollution, especially in the Huarisca and La Perla sectors. Meanwhile, in San Blas the benthic macroinvertebrate communities have a greater affinity for high EC, OD, and pH values. However, the results of BIOENV's best analysis show that turbidity is the variable that has the highest correlation value with the distribution of benthic macroinvertebrates.

## 5. Conclusion

The river Cunas constitutes an essential source of water for the diverse uses to the populations that settle in its basin. The quality of the water in the sampling sectors of the river reveals a progressive deterioration as anthropogenic activities increase as a result of the accelerated population growth and migration to urban areas in the region. The regular water quality in the Huarisca and La Perla sectors is due to the high loads of organic matter in the wastewater discharged into the river, the contribution of nutrients from detergents, and the drainage of fertilized agricultural soils. This condition of the river in these sectors would influence the composition of benthic macroinvertebrate communities. The presence of a higher number of individuals of the Chironomidae family, resilient to organic contamination, especially in the Huarisca and La Perla sectors, reveals the disturbance that the river has been experiencing.

## Acknowledgements

The authors express their gratitude to the National University of Central Peru for funding the study and to the Water Research Laboratory for allowing us to make use of the equipment and materials for this study.

## Conflict of interest

The authors declare that they have no conflict of interest.

## Authors' contributions

María Custodio developed the concept and design of the field study and performed the analysis of benthic macroinvertebrate communities, determination of thermotolerant coliforms, and writing of the manuscript. Heidi De la Cruz carried out the determination of the physical-chemical parameters in situ and in the laboratory. Richard Peñaloza carried out the water and sediment sampling, elaborated the location map of the study, and carried out the statistical analysis. All authors approved the final version prior to submission.

## Author details

María Custodio[1]*, Richard Peñaloza[2] and Heidi De La Cruz[1]

1 Universidad Nacional del Centro del Perú, Huancayo, Perú

2 Universidad Nacional Agraria La Molina, Lima, Perú

*Address all correspondence to: mcustodio@uncp.edu.pe

IntechOpen

# References

[1] Siidagyte E, Visinskiene G, Arbaciauskas K. Macroinvertebrate metrics and their integration for assessing the ecological status and biocontamination of Lithuanian lakes. Limnologica. 2013;**43**:308-318. DOI: 10.1016/j.limno.2013.01.003

[2] Miler O, Porst G, McGoff E, Pilotto F, Donohue L, Jurca T, et al. Morphological alterations of lake shores in Europe: A multimetric ecological assessment approach using benthic macroinvertebrates. Ecological Indicators. 2013;**34**:398-410. DOI: 10.1016/j.ecolind.2013.06.002

[3] Damanik-Ambarita M, Lock K, Boets P, Everaert G, Nguyen T, Forio E, et al. Ecological water quality analysis of the Guayas river basin (Ecuador) based on macroinvertebrates indices. Limnologica. 2016;**57**:27-59. DOI: 10.1016/j.limno.2016.01.001

[4] Forio E, Goethals M, Lock K, Asio V, Bande M, Thas O. Assessment and analysis of ecological quality, macroinvertebrate communities and diversity in rivers of a multifunctional tropical island. Ecological Indicators. 2018;**77**:228-238. DOI: 10.1016/j.envsoft.2017.11.025

[5] Ganguly I, Patnaik L, Nayak S. Macroinvertebrates and its impact in assessing water quality of riverine system: A case study of Mahanadi river, Cuttack, India. Journal of Natural and Applied Sciences. 2018;**10**:958-963. DOI: 10.31018/jans.v10i3.1817

[6] Forio E, Goethals M, Lock K, Asio V, Bande M, Thas O. Model-based analysis of the relationship between macroinvertebrate traits and environmental river conditions. Environmental Modelling and Software. 2018;**106**:57-67. DOI: 10.1016/j. envsoft.2017.11.025

[7] Scheibler E, Claps M, Roig-Juñent A. Temporal and altitudinal variations in benthic macroinvertebrate assemblages in an Andean river basin of Argentina. Journal of Limnology. 2014;**73**:76-92. DOI: 10.4081/jlimnol.2014.789

[8] Daneshvar F, Nejadhashemi A, Herman M, Abouali M. Response of benthic macroinvertebrate communities to climate change. Ecohydrology & Hydrobiology. 2017;**17**:63-72. DOI: 10.1016/j.ecohyd.2016.12.002

[9] Nguyen H, Forio M, Boets P, Lock K, Ambarita M, Suhareva N, et al. Threshold responses of macroinvertebrate communities to stream velocity in relation to hydropower dam: A case study from the Guayas River Basin (Ecuador). Water (Switzerland). 2018;**10**:4-17. DOI: 10.3390/w10091195

[10] National Water Authority. Water Resources Quality Monitoring Protocol National Water Authority. GreenFacts; 2009. p. 34

[11] APHA/AWWA/WEF. Standard Methods for the Examination of Water and Wastewater. Stand: Methods; 2012. p. 541

[12] Gabriels W, Lock K, De Pauw N, Goethals L, Goethals M. Multimetric Macroinvertebrate Index Flanders (MMIF) for biological assessment of rivers and lakes in Flanders (Belgium). Limnologica. 2010;**40**:199-207. DOI: 10.1016/j.limno.2009.10.001

[13] Huamantinco A, Ortiz W. Key of larval genera of Trichoptera (Insecta) of the Western slope of the Andes, Lima, Peru. Revista Peruana de Biología. 2010; **17**:75-80. DOI: 10.15381/rpb.v17i1.54

[14] Dutertre M, Hamon D, Chevalier C, Ehrhold A. The use of the relationships

between environmental factors and benthic macrofaunal distribution in the establishment of a baseline for coastal management. ICES Journal of Marine Science. 2013;**70**:294-308. DOI: 10.1093/icesjms/fss170

[15] Padovan A, Munksgaard N, Alvarez B, McGuinness K, Parry D, Gibb K. Trace metal concentrations in the tropical sponge Spheciospongia vagabunda at a sewage outfall: Synchrotron X-ray imaging reveals the micron-scale distribution of accumulated metals. Hydrobiologia. 2012;**687**:275-288. DOI: 10.1007/s10750-011-0916-9

[16] Ceschia C, Falace A, Warwick R. Biodiversity evaluation of the macroalgal flora of the Gulf of Trieste (Northern Adriatic Sea) using taxonomic distinctness indices. Hydrobiologia. 2007;**580**:43-56. DOI: 10.1007/s10750-006-0466-8

[17] Anderson M, Willis T. Canonical analysis of principal coordinates: A useful method of constrained ordination for ecology. Ecology. 2003;**84**:511-525. DOI: 10.1890/0012-9658(2003)084[0511:CAOPCA]2.0.CO;2

[18] Effendi H. Valuation of water quality status of Ciliwung River based on Pollution Index. In: A paper presented at 36th Annual Conference of the International Association for Impact Assessment (IAIA 16); 11-14 May 2016; Japan: Aichi-Nagoya

[19] MINEN. Supreme Decree No. 015-2015-MINEN—National Environmental Quality Standards for Water. Peru: Official Newspaper El Peruano; 2015. p. 3

[20] WHO. Guidelines for Drinking-Water Quality. Fourth ed. Geneva: WHO: World Health Organization; 2011. p. 564

[21] CCME. Canadian Water Quality Guidelines for the Protection of Aquatic Life [Internet]. 2007. Available from: http://www.ceqg-rcqe.ccme.ca/download/en/221

[22] Ayandiran T, Fawole O, Dahunsi S. Water quality assessment of bitumen polluted Oluwa River, South- Western Nigeria. Water Resources and Industry. 2018;**19**:13-24. DOI: 10.1016/j.wri.2017.12.002

[23] Seiler L, Helena E, Fernandes L, Martins F, Cesar P. Evaluation of hydrologic influence on water quality variation in a coastal lagoon through numerical modeling. Ecological Modelling. 2015;**314**:44-61. DOI: 10.1016/j.ecolmodel.2015.07.021

[24] Cony N, Ferrer N, Cáceres E. Evolution of the trophic state and phytoplankton structure of a Somero lake in the Pampean region: Laguna Sauce Grande (Province of Buenos Aires, Argentina). Biología Acuática. 2016;**30**:79-91

[25] Barakat A, El Baghdadi M, Rais J, Aghezzaf B, Slassi M. Assessment of spatial and seasonal water quality variation of Oum Er Rbia River (Morocco) using multivariate statistical techniques. International Soil and Water Conservation Research. 2016;**4**:284-292. DOI: 10.1016/j.iswcr.2016.11.002

[26] Paudel B, Weston N, O'Connor J, Sutter L, Velinsky D. Phosphorus dynamics in the water column and sediments of Barnegat Bay, New Jersey. Journal of Coastal Research. 2017;**78**:60-69. DOI: 10.2112/SI78-006.1

[27] Vamosi S, Silver C, Vamosi S. Macroinvertebrate community composition of temporary prairie wetlands: A preliminary test of the effect of macroinvertebrate community composition of temporary prairie wetlands: A preliminary test of the effect of rotational grazing. Wetlands. 2012;**32**:185-197. DOI: 10.1007/s13157-012-0268-x

[28] Riens J, Schwarz M, Hoback W. Aquatic macroinvertebrate communities and water quality at buffered and non-buffered wetland sites on federal waterfowl production areas in the Rainwater Basin, Nebraska. Wetlands. 2013;**33**:1025-1036. DOI: 10.1007/s13157-013-0460-7

[29] Miserendino M, Archangelsky M, Brand C, Epele L. Environmental changes and macroinvertebrate responses in Patagonian streams (Argentina) to ashfall from the Chaitén Volcano (May 2008). The Science of the Total Environment. 2012;**424**:202-212. DOI: 10.1016/j.scitotenv.2012.02.054

[30] Verdonschot R, Didderen K, Verdonschot P. Importance of habitat structure as a determinant of the taxonomic and functional composition of lentic macroinvertebrate assemblages. Limnologica—Ecology and Management of Inland Waters. 2012;**42**: 31-42. DOI: 10.1016/j. limno.2011.07.004

[31] Mykrä H, Saarinen T, Tolkkinen M, McFarland B, Hämäläinen H, Martinmäki K, et al. Spatial and temporal variability of diatom and macroinvertebrate communities: How representative are ecological classifications within a river system? Ecological Indicators. 2012;**18**:208-217. DOI: 10.1016/j.ecolind.2011.11.007

# Coral Reef Ecosystems

# Chapter 5

# Skeletons of Calcareous Benthic Hydroids (Medusozoa, Hydrozoa) under Ocean Acidification

*María A. Mendoza-Becerril, Crisalejandra Rivera-Perez and José Agüero*

## Abstract

The skeleton plays a vital role in the survival of aquatic invertebrates by separating and protecting them from a changing environment. In most of these organisms, calcium carbonate ($CaCO_3$) is the principal constituent of the skeleton, while in others, only a part of the skeleton is calcified, or $CaCO_3$ is integrated into an organic skeleton structure. The average pH of ocean surface waters has increased by 25% in acidity as a result of anthropogenic carbon dioxide ($CO_2$) emissions, which reduces carbonate ions ($CO_3^{2-}$) concentration, and saturation states ($\Omega$) of biologically critical $CaCO_3$ minerals like calcite, aragonite, and magnesian calcite (Mg-calcite), the fundamental building blocks for the skeletons of marine invertebrates. In this chapter, we discuss how ocean acidification (OA) affects particular species of benthic calcareous hydroids in order to bridge gaps and understand how these organisms can respond to a growing acidic ocean.

**Keywords:** biomineralization, Cnidaria, Hydractiniidae, Milleporidae, ocean acidification, skeleton, Stylasteridae

## 1. Introduction

Since the arrival of industrialization with the beginning of the British Industrial Revolution in 1750 to now, the accumulative concentration of carbon dioxide ($CO_2$) in the atmosphere through to the year 2019 has increased to 2340 ± 240 gigatonnes of $CO_2$ ($GtCO_2$), of which 25% has been sunk into the ocean [1, 2]. This human-induced sink of $CO_2$ in the ocean produces a chemical phenomenon called ocean acidification (OA) [3]. OA decreases seawater pH, the concentration of carbonate ions ($CO_3^{2-}$), and the saturation state ($\Omega$) of the three primary biogenic calcium carbonate ($CaCO_3$) minerals that occur in seawater and in shells and skeletons of calcifying organisms: calcite, aragonite, and magnesian calcite (Mg-calcite) [4].

Shells and skeletons of calcifying organisms play an essential role in their survival by separating and protecting them from a changing environment, as it happens with calcareous cnidarians [5, 6]. Within the phylum Cnidaria, only 17% of its extant species produce a calcareous skeleton through a process of biological transformation called biomineralization [7, 8]. The biomineralization process involves the selective extraction, transport, and uptake of biominerals from the environment in the function of their abundance and availability for their later incorporation into functional structures under strict biological control [8].

Of the 17% of the extant cnidarians with a calcareous skeleton, 14% is represented by members of the order Scleractinia (Cnidaria, Anthozoa), while the remaining 3% is made up of species belonging to the superorder "Anthoathecata" (Cnidaria, Hydrozoa) (**Figure 1**) [7]. In the class Anthozoa, the biomineralization process is the best known and most widely studied, being the opposite for the class Hydrozoa [9], although they are one of the main components of zoobenthic communities, significant contributors to the building of coral reefs (**Figure 2**) [10–12], and also some are essential in pelagic communities due to the presence of a medusa stage [10].

Calcareous hydroid families with a well-developed benthic polypoid stage are Milleporidae (hydrocorals, "fire corals" or millepores) with 15 species, Hydractiniidae (longhorn hydrozoans) with 4 species, and Stylasteridae (hydrocorals, lace corals, or stylasterids) with 320 species [7, 13]. These three families constitute a polyphyletic group and are commonly grouped as "calcified hydroids," "calcareous hydrocorals," or, simply, "hydrocorals"—terms that refer to hydroids that secrete a calcareous skeleton [14]. These calcareous structures can take the form of skeletons composed of individual spicules, spicule aggregates, or massive skeletons [15], and are responsible for providing protection and ion storage [6, 16, 17].

The calcareous skeleton of the cnidarians is always ectodermal in origin, and its mineralogy is composed exclusively of $CaCO_3$ [18]. In the calcareous species of the class Hydrozoa, their skeletons are composed of calcite, aragonite, or both (**Table 1**) [9, 19–23]. Calcite and aragonite are two of the six $CaCO_3$ polymorphs and are the most thermodynamically stable structures deposited extensively as biominerals [8]. In stylasterid species, for instance, the distribution of calcite or aragonite in their skeletons can be as follows: 100% calcite, 100% aragonite, primarily calcite with some aragonite, or primarily aragonite with some calcite [22]. When calcite and aragonite are present at the same time, the two polymorphs always occupy different anatomical sites [20]. Since the natural color of $CaCO_3$ is white [24], the broad spectrum of colors observed in the calcareous skeletons of hydrocorals is due to the presence of carotenoproteins, symbiotic dinoflagellates of the genus *Symbiodinium*, or by the presence of microboring or euendolithic microorganisms [25–27].

Phylogenetic analysis supports the independent origins of a calcified skeleton in Hydrozoa [9, 28, 29], and the distribution of $CaCO_3$ polymorphs in their skeletons is considered to have been produced by non-environmental causes [22]. However,

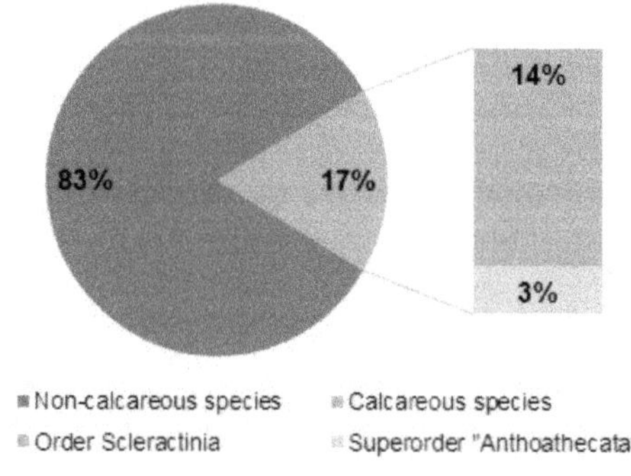


**Figure 1.**
*Worldwide inventory of non-calcareous and calcareous cnidarians. Own elaboration with WoRMS data [7].*

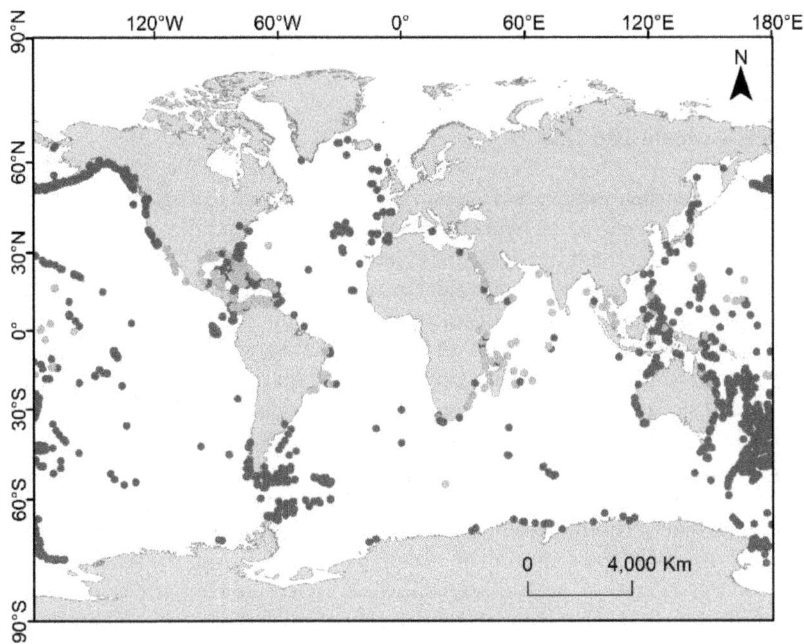

**Figure 2.**
*Worldwide hydrocorals and longhorn hydrozoans distribution. Orange dots, Milleporidae; green dots, Hydractiniidae; purple dots, Stylasteridae. Own elaboration with OBIS data [12].*

| Taxa | Type of skeletogenesis | Principal mineral |
|---|---|---|
| Subclass Hydroidolina | | |
| Superorder "Anthoathecata" | | |
| Order Capitata | | |
| Family Milleporidae | Modified spherulitic to trabecular | |
| *Millepora* spp. | | Aragonite |
| Order "Filifera" | | |
| Family Hydractiniidae | Spherulitic (with organic lamellae) | |
| *Distichozoon dens* | | Unknown |
| *Hydrocorella africana* | | Unknown |
| *Janaria mirabilis* | | Unknown |
| *Schuchertinia antonii* | | Unknown |
| Family Stylasteridae | Fully spherulitic or modified spherulitic to trabecular | |
| *Cheiloporidion pulvinatium* | | Primarily aragonite with some calcite |
| *Errina* sp. | | Primarily calcite with some aragonite |
| *Errinopsis* sp. | | Calcite |
| *Lepidopora* spp. | | Aragonite |

**Table 1.**
*Types of skeletogenesis and mineral composition of skeletons in calcareous Hydrozoa [19–22].*

the biomineralization process of these organisms is highly variable and strongly affected by environmental factors [30, 31] and substrate [32].

## 2. Skeletogenesis and OA

The biomineralization process is practically unknown to calcareous hydroids. Sorauf [21] summarizes some hypotheses about the biomineralization process of some Hydrozoa, and there has been no review about it to date. The basic structure is of the spherulitic growth of a principal mineral controlled by organic substrates to form pillars in which the spherulites are in part compartmentalized by a skeletal organic matrix (SOM), which forms an irregular matrix with compartments but does not form sheaths for individual crystal growth. In the class Hydrozoa exist three types of skeletogenesis, and the principal minerals involved are the $CaCO_3$ polymorphs aragonite or calcite (**Table 1**) [19–22].

In addition to biomineralization, $CaCO_3$ plays a significant role as second messenger to control exocytosis, cortical reactions in eggs, and muscle contraction [33]. In some hydractinids, $CaCO_3$ is required for larval motility [34], induction of metamorphosis [35], and secretion of adhesive material during the latter [34].

The biocrystallization, such as sclerotization, is derived from the ectoderm, which produces a SOM that controls the spacing of nucleation sites and limits the size or shape of spherulites [21, 36]. The organic secretions may be composed of peptides, proteins, proteoglycans, lipids, and polysaccharides, which, as a whole, are known as the template for mineralization [21, 37]. It is known that this template is involved in most, if not all, stages of biomineral formation, from transport, through nucleation and growth, to structure stabilization (**Figure 3**) [37].

According to an analysis of SOM homologs in cnidarians, including Hydrozoa, several proteins related to biomineralization were identified [38]. Extracellular

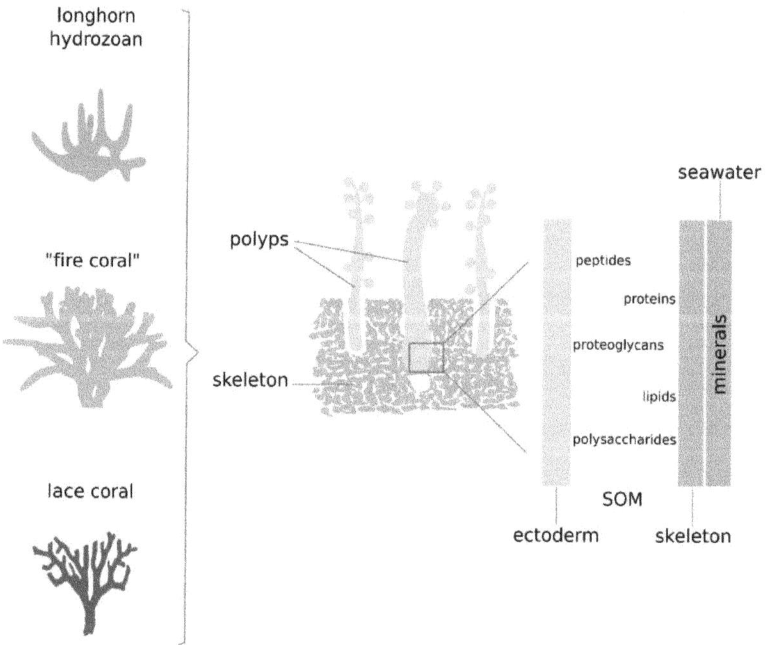


**Figure 3.**
*Schematic representation of the hypothetical skeletogenesis process on calcareous hydroids.*

adhesion proteins and carbonic anhydrases homologs were the most common proteins found (e.g., in *Millepora alcicornis*, *Millepora complanata*, and *Millepora squarrosa*). Homolog proteins include enzymes such as peptidase-1 and peptidase-2 as well as acidic proteins like SAARP-1, SAARP-2, acidic SOMP, CARP4, CARP5, Integrin-like and two SAARP-like proteins; those proteins are involved in calcite formation [39]. Two galaxin ortholog proteins (Galaxin and Galaxin-2) [38] have been fully characterized by the calcifying matrix of scleractinian corals [40]. More interestingly, carbonic anhydrases, which are known to precipitate $CaCO_3$ in different calcareous organisms [41], have been identified in Hydrozoa species, CruCA-4, and Putative CA [38]. Finally, in contrast to scleractinian corals, Hydrozoa species did not show small cysteine-rich proteins (SCRiPs) [38], whose function in corals is still unclear.

In some calcareous hydroids, a progressive capability to produce a similar SOM to that of scleractinian corals has been observed, with individual control of crystal growth [21]. Also, the calcification process of stylasterid and millepore species has been compared with that of scleractinian corals [42]. This calcification process includes uptake and transport of materials, production of organic secretion, the formation of tissue cavities where calcification may take place, and the deposition of $CaCO_3$; these processes may be influenced differently by environmental conditions, and be affected by OA [43, 44].

About the biochemical process underlying the response of hydrocoral *M. alcicornis* in acidified waters, it has been found that the calcification process in the hydrocoral was not affected by a wide range of seawater pH (8.1–7.5) under experimental conditions [30]. Besides, is mentioned that the Ca-ATPase plays an essential role in the biomineralization as maintenance a steady-state net calcification rate in the hydrocoral, especially under scenarios of moderate (pH 7.8) and intermediate (pH 7.5) acidification of seawater, but under a scenario of severe acidification (pH 7.2) of seawater, the hydrocoral is not able to maintain a steady-state net calcification rate [30]. On the other hand, physiologically, the exposure to seawater acidification induces oxidative stress with consequent oxidative damage to lipids and proteins, which could compromise hydrocoral health [45]. However, a reduction in the calcification process was not observed in *Millepora platyphylla* despite having been exposed to OA conditions [46].

Some effects in other calcareous organisms, for instance, anthozoans, sea urchins, and mollusks by OA are: slowdown of their calcification rates; changes in gene expression consistent with metabolic suppression; increased oxidative stress; potential effect on symbiotic zooxanthellae; decrease in matrix proteins; reduction of carbonic anhydrase protein; increased calcite growth; structural disorientation of calcite crystals; fragile skeletons that reduce protection from predators and changing environments, affect the expression of the gene encoding Ca-ATPase enzymes and the enzymatic activity itself [30, 44, 47–49].

OA not only affects the skeleton of the calcareous hydroids, but it can also affect the other phases of its life cycle, for instance, the medusa stage of millepores, since it has been recently recorded that cubomedusae suffer from higher mortality when subjected to OA conditions (pH 7.5) [50].


## 3. Implications, threats, and consequences of OA

The response of hydrocorals to the changes they face in their environment remains unknown, especially how they are affected by anthropogenic activities such as the increase in the concentration of $CO_2$ in the atmosphere, causing an increase in sea surface temperature (SST) and a decrease in seawater pH. The chemistry of

OA is better understood from their implications for calcifying marine fauna and their hosts or associations. Skeletons of hydrocorals and longhorn hydrozoans are known to host abundant and diverse symbiotic organisms, for instance, with photosynthetic dinoflagellates (generally referred to as zooxanthellae), and maintain associations with micro and macroboring organisms, and grazers. The microboring organisms (MIO) include cyanobacteria, green and red algae, fungi, and lichens [51]. The macroboring organisms (MAO) comprise ascidians and sponges [42], while in the grazers encompass echinoderms, mollusks, polychaeta, crustaceans, and fish [42, 52].

Of the three families of extant calcareous hydroids, only "fire corals" have a symbiotic relationship with zooxanthellae [42]. The zooxanthellae are essential for the "fire corals" to achieve their calcification process, keep their rate of calcification constant, as well as speed up a calcareous deposition in the function of the environmental conditions [43]. Loss of this association from hydrocoral tissue is responsible for the white color observed, aptly named bleaching [53]. When "fire corals" experiment stress occurs bleaching, or the paling zooxanthellate decline and the concentration of pigments within the zooxanthellae fall, where each zooxanthella may lose 50–80% of its photosynthetic pigments [54]. The stress can be induced by a plethora of factors, singly or in combination, and among them we have: anomalously low and high temperature, solar radiation, subaerial exposure, sedimentation, freshwater dilution, inorganic nutrients, high concentrations of xenobiotics, presence of pathogens such as protozoan and bacterium, OA, among others [54, 55]. Recently, it has been observed that hydrocorals can select their symbionts zooxanthellae, depending on environmental conditions, which can confer an advantage on how to face ongoing human-driven climate change [56].

The mechanism underlying the observed bleaching response was not explicitly investigated, some hypotheses are that changes in seawater chemistry influence bleaching thresholds by altering the functioning of the carbon-concentrating mechanism (CCM), photoprotective mechanisms (such as photorespiration for instance), or direct impacts of acidosis; therefore, the acidification effects on coral bleaching are uncertain and review of other aspects, for instance, levels of the other abiotic factors such as light and nutrients, photoacclimation and photoprotection responses, molecular genetics, as well as studies that imply the understanding of integral processes about host-algae are recommended to understand the role that zooxanthellae may play in the ability of corals to cope with these anthropogenic changes in the ocean [53, 57, 58].

The MIO distribution within the skeletons occurs through contact with the substrate of settlement as MIO already colonizes it, and their colonization occurs early in the development of the corals and expands at slower rates than the hydrocoral growth [27]. Since stylasterid corals do not host zooxanthellae, such an arrangement may be beneficial throughout the life of the coral, despite some losses to its skeleton density due to dissolution by MIO; moreover, the boring microflora within corals have a mutualistic relationship, helping corals survive better during bleaching events, because these MIO may satisfy the nitrogen quantities required by live hydrocorals for their balanced growth, also considering that MIO are the major primary producers and agents of microbioerosion dissolving large quantities of $CaCO_3$ with a potential in buffering seawater [59].

Micro and macrobioerosion under undisturbed natural conditions are essential mechanisms in $CaCO_3$ recycling; however, these bioerosion processes can proceed faster if OA weakens the substrate, also facilitating in this way the bioerosion by grazers [60]. Furthermore, OA does not affect the siliceous sponges as directly as other marine taxa, which are heavily dependent on $CaCO_3$ at various life history stages like cnidarians, mollusks, and many crustaceans species with tiny pelagic larval forms [61]. These siliceous sponges represent a threat when settling on calcareous substrates by the process of weakening the skeleton by their bioeroder action; nevertheless, thermal stress appears to weaken calcifiers more strongly than bioeroding sponges [62].

Other impacts include shifts in competitive interactions with non-reef builders such as macroalgae, sponges, soft corals, ascidians, and corallimorpharians; the competition impacts the recruitment, growth, and mortality of coral organisms [63].

## 4. Conclusion

This review of current literature concerning the effects of OA on hydrocorals and longhorn hydrozoans and their proposed mechanisms shows that targets are numerous, and therefore it is difficult today to give a conclusion. Besides, several of the findings correspond to anthozoans and specific areas or under laboratory or modeling conditions. On the other hand, it has been shown that each species has a different response, some are more sensitive than others, and some show strategies to survive under conditions of anthropogenic climate change. As proposed by Luz [45], further studies that use metabolomics and proteomics techniques are necessary to help identify different response pathways in hydrocorals exposed to acidic conditions.

## Acknowledgements

This work was supported by Medusozoa Mexico (https://medusozoamexico.com.mx/). Special thanks are due to Mariae C. Estrada-González for her help with the design of **Figure 2**.

## Conflict of interest

The authors declare no conflict of interest.

## Author details

María A. Mendoza-Becerril[1*], Crisalejandra Rivera-Perez[1] and José Agüero[2]

1 CONACyT-Northwestern Center for Biological Research, Instituto Politécnico Nacional, La Paz, Mexico

2 Medusozoa Mexico, La Paz, Mexico

*Address all correspondence to: m_angelesmb@hotmail.com

IntechOpen

## References

[1] O'Brien PK. Deconstructing the British industrial revolution as a conjuncture and paradigm for global economic history. In: Horn J, Rosenband LN, Smith MR, editors. Reconceptualizing the Industrial Revolution. United States of America: MIT Press; 2010. pp. 21-46. Available from: http://www.jstor.org/stable/j.ctt5hhgdm.6

[2] Friedlingstein P, Jones MW, O'Sullivan M, Andrew RM, Hauck J, Peters GP, et al. Global carbon budget 2019. Earth System Science Data. 2019;**11**(4):1783-1838. Available from: https://www.earth-syst-sci-data.net/11/1783/2019/

[3] Caldeira K, Wickett ME. Anthropogenic carbon and ocean pH. Nature. 2003;**425**(6956):365. DOI: 10.1038/425365a

[4] Gattuso J-P, Hansson L. Ocean acidification: Background and history. In: Gattuso J-P, Hansoon L, editors. Ocean Acidification. Pondicherry: Oxford University Press; 2011. pp. 1-20

[5] Watabe N. Shell. In: Bereiter-Hahn J, Matoltsy AG, Richards KS, editors. Biology of the Integument 1: Invertebrates. 1st ed. Berlin, Heidelberg: Springer-Verlag Berlin, Heidelberg; 1984. pp. 448-485. DOI: 10.1007/978-3-642-51593-4_25

[6] Tidball JG. Cnidaria: Secreted surface. In: Bereiter-Hahn J, Matoltsy AG, Richards KS, editors. Biology of the Integument: Invertebrates. Berlin, Heidelberg: Springer Berlin Heidelberg; 1984. pp. 69-78. DOI: 10.1007/978-3-642-51593-4_7

[7] WoRMS. Cnidaria; 2019 [cited 01 December 2019]. Available from: http://www.marinespecies.org/aphia.php?p=taxdetails&id=1267

[8] Mann S. In: Mann S, editor. Biomineralization: Principles and Concepts in Bioinorganic Materials Chemistry. 1st ed. New York, United States: Oxford University Press; 2001. 198 p. (Oxford Chemistry Masters)

[9] Miglietta MP, McNally L, Cunningham CW. Evolution of calcium-carbonate skeletons in the Hydractiniidae. Integrative and Comparative Biology. 2010;**50**(3):428-435. DOI: 10.1093/icb/icq102

[10] Di Camillo CG, Bavestrello G, Cerrano C, Gravili C, Piraino S, Puce S, et al. Hydroids (Cnidaria, Hydrozoa): A neglected component of animal forests. In: Rossi S, Bramanti L, Gori A, Covadonga O, editors. Marine Animal Forests: The Ecology of Benthic Biodiversity Hotspots. 1st ed. Cham, Switzerland: Springer International Publishing; 2017. pp. 397-427. DOI: 10.1007/978-3-319-21012-4_11

[11] Weber JN, Woodhead PMJ. Stable isotope ratio variations in non-scleractinian coelenterate carbonates as a function of temperature. Marine Biology. 1972;**15**(4):293-297. DOI: 10.1007/BF00401388

[12] OBIS. Ocean Biogeographic Information System. Intergovernmental Oceanographic Commission of UNESCO; 2019 [cited 02 November 2019]. Available from: http://www.iobis.org

[13] Cairns SD. Deep-water corals: An overview with special reference to diversity and distribution of deep-water scleractinian corals. Bulletin of Marine Science. 2007;**81**(3):311-322. Available from: https://www.ingentaconnect.com/search/article?option1=tka&value1=deep.waters+corals.+An+overview+with+special&pageSize=10&index=1

[14] Amaral FMD, Steiner AQ, Broadhurst MK, Cairns SD. An overview

of the shallow-water calcified hydroids from Brazil (Hydrozoa: Cnidaria), including the description of a new species. Zootaxa. 2008;**1930**(1):56-68. DOI: 10.11646/zootaxa.1930.1.4

[15] Lowenstam HA, Weiner S. In: Lowenstam HA, Weiner S, editors. On Biomineralization. 1st ed. USA: Oxford University Press; 1989. 324 p

[16] Chapman DM, Histology C. In: Muscatine L, Lenhoff H, editors. Coelenterate Biology Review and New Perspectives. 1st ed. UK: Academic Press; 1974. pp. 1-92. Available from: http://www.sciencedirect.com/science/article/pii/B9780125121507500062

[17] Harrison FW, Ruppert EE. In: Harrison FW, Ruppert EE, editors. Microscopic Anatomy of Invertebrates, Volume 2, Placozoa, Porifera, Cnidaria, and Ctenophora. 1st ed. United States of America: Wiley; 1991. 456 p. (Microscopic Anatomy of Invertebrates)

[18] Cuif J-P. Calcification in the Cnidaria through time: An overview of their skeletal patterns from individual to evolutionary viewpoints. In: Goffredo S, Dubinsky Z, editors. The Cnidaria, Past, Present and Future: The World of Medusa and her Sisters. Cham: Springer International Publishing; 2016. pp. 163-179. DOI: 10.1007/978-3-319-31305-4_11

[19] Moore RC, Wells JW, Harrington HJ, Hill D, Boschma H, Hyman LH, et al. Coelenterata. In: Moore RC, editor. Treatise on Invertebrate Paleontology. USA: Geological Society of America, Incorporated; 1956. pp. 1-106

[20] Lowenstam HA. Coexisting calcites and aragonites from skeletal carbonates of marine organisms and their strontium and magnesium contents. In: Miyake Y, Koyama T, editors. Recent Researches in the Fields of Hydrosphere, Atmosphere and Nuclear Geochemistry. 1st ed. Tokyo, Japan: Editorial Committee for Sugawara Volume; 1964. pp. 373-404

[21] Sorauf JE. Biomineralization, structure and diagenesis of the coelenterate skeleton. Acta Palaeontologica Polonica. 1980;**25**(3-4):327-343

[22] Cairns SD, Macintyre IG. Phylogenetic implications of calcium Carbonate mineralogy in the Stylasteridae (Cnidaria: Hydrozoa). PALAIOS. 1992;7(1):96-107. Available from: http://www.jstor.org/stable/3514799

[23] Cairns SD. In: Cairns SD, editor. Stylasteridae (Cnidaria: Hydrozoa: Anthoathecata) of the New Caledonian Region. Paris: Muséum national d'Histoire naturelle; 2015. 362 p. (Mémoires du Muséum national d'histoire naturelle)

[24] Al Omari MMH, Rashid IS, Qinna NA, Jaber AM, Badwan AA. Calcium carbonate. In: Brittain HG, editor. Profiles of Drug Substances, Excipients and Related Methodology. 1st ed. USA: Academic Press; 2016. pp. 31-132. Available from: http://www.sciencedirect.com/science/article/pii/S1871512515000229

[25] Rønneberg H, Fox DL, Liaaen-Jensen S. Animal carotenoids—Carotenoproteins from hydrocorals. Comparative Biochemistry and Physiology Part B: Comparative Biochemistry. 1979;**64**(4):407-408. Available from: http://www.sciencedirect.com/science/article/pii/030504917990292X

[26] LaJeunesse T. Diversity and community structure of symbiotic dinoflagellates from Caribbean coral reefs. Marine Biology. 2002;**141**(2):387-400. DOI: 10.1007/s00227-002-0829-2

[27] Pica D, Tribollet A, Golubic S, Bo M, Di CCG, Bavestrello G, et al. Microboring organisms in living stylasterid corals (Cnidaria, Hydrozoa). Marine Biology Research.

2016;**12**(6):573-582. DOI: 10.1080/17451000.2016.1169298

[28] Lindner A, Cairns SD, Cunningham CW. From offshore to onshore: Multiple origins of shallow-water corals from Deep-Sea ancestors. PLoS One. 2008;**3**(6):e2429. DOI: 10.1371/journal.pone.0002429

[29] Mendoza-Becerril MA, Jaimes-Becerra AJ, Collins AG, Marques AC. Phylogeny and morphological evolution of the so-called bougainvilliids (Hydrozoa, Hydroidolina). Zoologica Scripta. 2018;**47**(5):608-622. Available from: https://onlinelibrary.wiley.com/doi/abs/10.1111/zsc.12291

[30] Marangoni de BLF, Calderon EN, Marques JA, Duarte GAS, Pereira CM, e Castro CB, et al. Effects of CO2-driven acidification of seawater on the calcification process in the calcareous hydrozoan *Millepora alcicornis* (Linnaeus, 1758). Coral Reefs. 2017;**36**(4):1133-1141. DOI: 10.1007/s00338-017-1605-6

[31] Olguín-López N, Hérnandez-Elizárraga VH, Hernández-Matehuala R, Cruz-Hernández A, Guevara-González R, Caballero-Pérez J, et al. Impact of El Niño-southern oscillation 2015-2016 on the soluble proteomic profile and cytolytic activity of *Millepora alcicornis* ("fire coral") from the Mexican Caribbean. PeerJ. 2019;**7**:e6593. DOI: 10.7717/peerj.6593

[32] Vago R, Pasternak G, Itzhak D. Aluminum metallic substrate induce colossal biomineralization of the calcareous hydrocoral *Millepora dichotoma*. Journal of Materials Science Letters. 2001;**20**(11):1049-1050. DOI: 10.1023/A:1010972727293

[33] Alberts B, Johnson A, Lewis J, Morgan D, Raff M, Roberts K, et al. Myosin and actin. In: Alberts B, Johnson A, Lewis J, Morgan D, Raff M, Roberts K, et al., editors. Molecular Biology of the Cell. 6th ed. United States of America: Garland Science; 2014. pp. 915-925

[34] Dandar-Roh AM, Rogers-Lowery CL, Zellmann E, Thomas MB. Ultrastructure of the calcium-sequestering gastrodermal cell in the hydroid *Hydractinia symbiolongicarpus* (Cnidaria, Hydrozoa). Journal of Morphology. 2004;**260**(2):255-270. DOI: 10.1002/jmor.10220

[35] Thomas MB, Edwards NC, Ball BE, McCauley DW. Comparison of metamorphic induction in hydroids. Invertebrate Biology. 1997;**116**(4):277-285. Available from: www.jstor.org/stable/3226859

[36] Knight DP. Sclerotization of the perisarc of the calyptoblastic hydroid, *Laomedea flexuosa*: 1. The identification and localization of dopamine in the hydroid. Tissue & Cell. 1970;**2**(3):467-477. Available from: http://www.sciencedirect.com/science/article/pii/S0040816670800453

[37] Ehrlich H. Chitin and collagen as universal and alternative templates in biomineralization. International Geology Review. 2010;**52**(7-8):661-699. DOI: 10.1080/00206811003679521

[38] Conci N, Gert W, Sergio V. New non-Bilaterian transcriptomes provide novel insights into the evolution of Coral Skeletomes. Genome Biology and Evolution. 2019;**11**(11):3068-3081. DOI: 10.1093/gbe/evz199

[39] Rahman MA, Tamotsu O. Aspartic acid-rich proteins in insoluble organic matrix play a key role in the growth of calcitic sclerites in alcyonarian coral. Chinese Journal of Biotechnology. 2008;**24**(12):2127-2128. DOI: 10.1016/S1872-2075(09)60003-0

[40] Fukuda I, Ooki S, Fujita T, Murayama E, Nagasawa H, Isa Y, et al. Molecular cloning of a cDNA encoding a

soluble protein in the coral exoskeleton. Biochemical and Biophysical Research Communications. 2003;**304**(1):11-17. Available from: http://www.sciencedirect.com/science/article/pii/S0006291X03005278

[41] Zoccola D, Innocenti A, Bertucci A, Tambutté E, Supuran CT, Tambutté S. Coral carbonic anhydrases: Regulation by ocean acidification. Marine Drugs. 2016;**14**(109):1-11. DOI: 10.3390/md14060109

[42] Lewis JBBT. Biology and ecology of the hydrocoral *Millepora* on Coral reefs. In: Southward AJ, Young CM, Fuiman LA, editors. Advances in Marine Biology. UK: Elsevier; 2006. pp. 1-55. Available from: http://www.sciencedirect.com/science/article/pii/S0065288105500014

[43] Strömgren T. Skeleton growth of the hydrocoral *Millepora complanata* Lamarck in relation to light. Limnology and Oceanography. 1976;**21**(1):156-160. DOI: 10.4319/lo.1976.21.1.0156

[44] Erez J, Reynaud S, Silverman J, Schneider K, Allemand D. Coral calcification under ocean acidification and global change. In: Dubinsky Z, Stambler N, editors. Coral Reefs: An Ecosystem in Transition. Dordrecht: Springer Netherlands; 2011. pp. 151-176. DOI: 10.1007/978-94-007-0114-4_10

[45] Luz DC, Zebral YD, Klein RD, Marques JA, Marangoni LF de B, Pereira CM, et al. Oxidative stress in the hydrocoral *Millepora alcicornis* exposed to CO2-driven seawater acidification. Coral Reefs. 2018;**37**(2):571-579. DOI: 10.1007/s00338-018-1681-2

[46] Brown D, Edmunds PJ. Differences in the responses of three scleractinians and the hydrocoral *Millepora platyphylla* to ocean acidification. Marine Biology. 2016;**163**(3):62. DOI: 10.1007/s00227-016-2837-7

[47] Orr JC, Fabry VJ, Aumont O, Bopp L, Doney SC, Feely RA, et al. Anthropogenic Ocean acidification over the twenty-first century and its impact on calcifying organisms. Nature. 2005;**437**(7059):681-686. DOI: 10.1038/nature04095

[48] Doney SC, Fabry VJ, Feely RA, Kleypas JA. Ocean acidification: The other CO2 problem. Annual Review of Marine Science. 2009;**1**(1):169-192. DOI: 10.1146/annurev.marine.010908.163834

[49] Fitzer SC, Phoenix VR, Cusack M, Kamenos NA. Ocean acidification impacts mussel control on biomineralisation. Scientific Reports. 2014;**4**(1):6218. DOI: 10.1038/srep06218

[50] Chuard PJC, Johnson MD, Guichard F. Ocean acidification causes mortality in the medusa stage of the cubozoan *Carybdea xaymacana*. Scientific Reports. 2019;**9**(1):5622. DOI: 10.1038/s41598-019-42121-0

[51] Tribollet A, Pica D, Puce S, Radtke G, Campbell SE, Golubic S. Euendolithic Conchocelis stage (Bangiales, Rhodophyta) in the skeletons of live stylasterid reef corals. Marine Biodiversity. 2018;**48**(4):1855-1862. DOI: 10.1007/s12526-017-0684-5

[52] Carpenter RC. Invertebrate predators and grazers. In: Birkeland C, editor. Life and Death of Coral Reefs. 1st ed. New York: Springer US; 1997. pp. 198-229

[53] Kirk Nathan L, Weis VM. Animal–symbiodinium symbioses: Foundations of coral reef ecosystems. In: Hurst CJ, editor. The Mechanistic Benefits of Microbial Symbionts. 1st ed. Cham, Zwitzerland: Springer International Publishing; 2016. pp. 269-294. DOI: 10.1007/978-3-319-28068-4_10

[54] Glynn PW. Coral reef bleaching: Facts, hypotheses and implications. Global Change Biology. 1996;**2**(6):495-509. DOI: 10.1111/j.1365-2486.1996.tb00063.x

[55] Eakin CM, Lough JM, Heron SF. Climate variability and change: Monitoring data and evidence for increased coral bleaching stress. In: van Oppen MJH, Lough JM, editors. Coral Bleaching: Patterns, Processes, Causes and Consequences. 1st ed. Berlin, Heidelberg: Springer Berlin Heidelberg; 2009. pp. 41-67. DOI: 10.1007/978-3-540-69775-6_4

[56] Rodríguez L, López C, Casado-Amezua P, Ruiz-Ramos DV, Martínez B, Banaszak A, et al. Genetic relationships of the hydrocoral *Millepora alcicornis* and its symbionts within and between locations across the Atlantic. Coral Reefs. 2019;**38**(2):255-268. DOI: 10.1007/s00338-019-01772-1

[57] Lesser MP. Coral bleaching: Causes and mechanisms. In: Dubinsky Z, Stambler N, editors. Coral Reefs: An Ecosystem in Transition. 1st ed. Dordrecht: Springer Netherlands; 2011. pp. 405-419. DOI: 10.1007/978-94-007-0114-4_23

[58] Albright R. Ocean acidification and coral bleaching. In: van Oppen MJH, and Lough JM, editors. Coral Bleaching: Patterns, Processes, Causes and Consequences. Cham, Switzerland: Springer International Publishing; 2018. pp. 295-323. DOI: 10.1007/978-3-319-75393-5_12

[59] Tribollet A. The boring microflora in modern coral reef ecosystems: A review of its roles. In: Wisshak M, Tapanila L, editors. Current Developments in Bioerosion. 1st ed. Berlin, Heidelberg: Springer Berlin Heidelberg; 2008. pp. 67-94. DOI: 10.1007/978-3-540-77598-0_4

[60] Schönberg CHL, Fang JKH, Carreiro-Silva M, Tribollet A, Wisshak M. Bioerosion: The other ocean acidification problem. ICES Journal of Marine Science. 2017;**74**(4):895-925. DOI: 10.1093/icesjms/fsw254

[61] Conway KW, Whitney F, Leys SP, Barrie JV, Krautter M. Sponge reefs of the British Columbia, Canada Coast: Impacts of climate change and ocean acidification. In: Carballo JL, Bell JJ, editors. Climate Change, Ocean Acidification and Sponges: Impacts Across Multiple Levels of Organization. 1st ed. Cham, Switzerland: Springer International Publishing; 2017. pp. 429-445. DOI: 10.1007/978-3-319-59008-0_10

[62] Schönberg CHL, Fang JK-H, Carballo JL. Bioeroding sponges and the future of coral reefs. In: Carballo JL, Bell JJ, editors. Climate Change, Ocean Acidification and Sponges: Impacts Across Multiple Levels of Organization. 1st ed. Cham, Switzerland: Springer International Publishing; 2017. pp. 179-372. DOI: 10.1007/978-3-319-59008-0_7

[63] Chadwick Nanette E, Morrow KM. Competition among sessile organisms on coral reefs. In: Dubinsky Z, Stambler N, editors. Coral Reefs: An Ecosystem in Transition. 1st ed. Dordrecht: Springer Netherlands; 2011. pp. 347-371. DOI: 10.1007/978-94-007-0114-4_20

# Mesophotic and Deep-Sea Vulnerable Coral Habitats of the Mediterranean Sea: Overview and Conservation Perspectives

*Giovanni Chimienti, Francesco Mastrototaro and Gianfranco D'Onghia*

## Abstract

Although the different communities distributed in the mesophotic and deep waters play a fundamental role in the functioning of the marine ecosystems, the assessment of their global distribution is still far to be achieved. This is also true in the Mediterranean Sea, where exploration technologies are revealing a large diversity of unknown communities structured totally or partially by corals, which represent vulnerable marine ecosystems (VMEs) according to FAO's guidelines. This chapter aims to define and describe the main coral habitats of the mesophotic and aphotic zones of the Mediterranean, such as coralligenous formations, cold-water coral frameworks, coral forests and sea pen fields. The role of these habitats in providing benefit for a variety of invertebrates and fishes as well as a suite of ecosystem goods and services is highlighted. Fishing is one of the main anthropogenic impacts affecting Mediterranean coral habitats, and the current conservation measures are often ineffective. Ongoing attempts and future solutions aiming at the conservation of these VMEs are here discussed, including the fishing restriction in strategic areas characterized by lush coral communities, the implementation of controls against illegal fishery, the development of encounter protocols for vulnerable species and the use of observers onboard.

**Keywords:** vulnerable marine ecosystems, corals, coralligenous, cold-water corals, sea pens, conservation, Mediterranean

## 1. Introduction

The Mediterranean Sea covers only 0.7% of the world's ocean surface, but it hosts about 8% of the global marine biodiversity [1]. Despite the presence of many anthropic impacts, a high diversity of benthic habitats is present from the coastal zones to the deep-sea bottom of this semi-enclosed basin, representing a worldwide hot spot of biodiversity [2]. Some of these habitats are built by one or few ecosystem engineer species that create a biogenic habitat suitable for a variety of associated species, including endangered and protected ones, as well as species of high commercial value. This is the case of the *Posidonia oceanica* meadows [3], *Cystoseira* spp. canopies [4], rhodolith beds [5], as well as the so-called marine bioconstructions [6] and animal forests *sensu lato* [7].

Knowledge, protection and the management of many of these habitats are still scarce and fragmentary, except for the *P. oceanica* meadows, which have received a substantial attention in the past and are now all included in Sites of Community Importance thanks to the Habitat Directive [8]. Several other biogenic habitats are worthy of protection, most of them occurring in the mesophotic and in the deep-sea zones. These habitats are more difficult to be explored and studied with respect to coastal environments. In these zones, the sunlight decreases with the depth until it disappears, and the animal life develops with emerging shapes and structures. In fact, the light is one of the main abiotic drivers determining the distribution of benthic organisms, together with oxygen concentration, water movement, temperature, pressure, sedimentation rates, substrate type and geographical area [9]. According to the penetration of the sunlight and its effects on the macrobenthic communities, the marine environment is conventionally divided into: euphotic zone, where the irradiance is strong enough to allow the development of seagrass; mesophotic or twilight zone, from the limit of the seagrass to the limit of presence of algae (loss of net productivity at level of irradiance <1%) [10]; and aphotic zone or deep-sea, where the light is absent. Although there are some indicative and non-univocal depth ranges for each zone (e.g., in the Mediterranean Sea: euphotic zone above 50 m depth; mesophotic zone ca. 50–150 m depth; and aphotic zone below 150–200 m depth), they can vary according to the water transparency and other physical factors that affect light penetration [11].

Corals are among the main habitat formers of the Mediterranean mesophotic and aphotic zones, constituting vulnerable marine ecosystems (VMEs) intended as those populations, communities or habitats that may be vulnerable to impacts from fishing activities [12]. Recently, the General Fishery Commission for the Mediterranean (GFCM) defined Mediterranean VME indicators taxa, habitats and features [13–15]. Corals, together with sponges, echinoderms, molluscs and other benthic organisms, play a significant role in the formation of VMEs. Moreover, most of coral taxa are included in relevant lists of protected species, such as the Red List drawn by the International Union for the Conservation of Nature (IUCN) [16]. Their vulnerability is due to their uniqueness or rarity, functional significance, fragility, structural complexity, as well as their life-history traits which make their recovery difficult (e.g., slow growth rate, late age of maturity, low or unpredictable recruitment, long life expectancy). For these reasons, several Mediterranean mesophotic and deep-sea coral habitats are particularly vulnerable to the impacts of different fishing gears, deserving *ad hoc* management measures [17–22].

This chapter defines and describes the main vulnerable ecosystems of the Mediterranean mesophotic and aphotic zones that are totally or partially based on the habitat-forming activity of corals. Their features, together with their ecological importance and ecosystem goods and services that they provide are highlighted, as well as the need of proper conservation measures.

## 2. Mesophotic and deep-sea vulnerable marine ecosystems

Marine bioconstructions are among the most important Mediterranean habitats below 50 m depth and highly vulnerable to fishing practices [6]. The term includes all those biogenic structures built-up by organisms growing one on the other, generation after generation. Most of this secondary substrate is constructed by species that accumulate calcium carbonate, constituting peculiar structures as coralligenous outcrops and Cold-Water Coral (CWC) frameworks [6, 23]. Further sensitive habitats of the mesophotic and aphotic zones are mostly represented by the so-called coral forests (i.e., communities structured by one or few coral species characterized by a typical arborescent morphology) [7] and the fields formed by sea pens [24].

## 2.1 Coralligenous

The term coralligenous groups a variety of temperate biogenic formations built-up by a mixed community of coralline red algae, corals, bryozoans, sponges, serpulids and molluscs (**Figure 1a**). Coralligenous communities thrive in the

**Figure 1.**
*Some of the mesophotic and deep-sea vulnerable habitats structured by corals in the Mediterranean Sea.*
*(a) Coralligenous formation with the dominant presence of the red gorgonian* Paramuricea clavata *(photo:*
*A. Sorci; Tremiti Islands, Adriatic Sea); (b) CWC framework built up by the scleractinians* Madrepora
oculata, Lophelia pertusa *and* Desmophyllum dianthus *(Santa Maria di Leuca, Ionian Sea); (c) forest*
*of the black coral* Antipathella subpinnata *(Tremiti Islands, Adriatic Sea); (d) forest of the gorgonian*
Callogorgia verticillata *on a deep hard bottom (Montenegro, Adriatic Sea); (e) forest of the bamboo*
*coral* Isidella elongata *on a deep muddy bottom (off Ibiza; Balearic Sea); and (f) field of the red sea pen*
Pennatula rubra *(Punta Alice, Ionian Sea).*

Mediterranean Sea from shallow waters (15–20 m of depth) up to the limit of the mesophotic zone [25, 26], and their existence is known since ancient time. The term "coralligène" was used for the first time by Marion [27] to describe the biogenic outcrops present between 30 and 70 m in depth in the Gulf of Marseille, called *broundo* by the local fishermen. Although the term *coralligène* literally means "coral producer," corals are not often the dominant component of the community and it is likely that Marion used the term "coraux" with the wide common meaning of his time, referring not strictly to scleractinians but to all those organisms that accumulate calcareous deposits such as corals, bryozoans and coralline algae. Afterward, also Pruvot [28–30] used the term *coralligène* to describe similar formations off Banyuls-sur-Mer (Gulf of Lion), and from the end of the nineteenth century, the coralligenous has been included in the bionomic description of the Mediterranean seabed [31].

Encrusting red algae represent a relevant component of the coralligenous formations, becoming less present with the decreasing of the light and disappearing in the aphotic zone, where the animal component dominates the bioconstructions. Locally, one or few species can be particularly abundant, representing the so-called *facies* [31], such as those of scleractinians, bryozoans or oysters. Arborescent anthozoans, such as black corals and alcyonaceans, can use the bioconstructions as secondary hard substratum to settle and form coral forests, increasing the three-dimensionality of this habitat (**Figure 1a** and **c**).

The coralligenous habitat represents a hot spot of biodiversity whose importance is comparable to that of the coral reefs in tropical ecosystems. In [25], the first estimate of the number of species associated with coralligenous formations is made, with about 1670 species. However, this is probably an underestimated number, because the complex structure of coralligenous assemblages and their highly diverse composition suggest that they probably host more species than any other Mediterranean habitats. This associated biodiversity includes species of conservation interest, as well as crustaceans and fish of high commercial value [32]. Thanks to the activity of its bioconstructors, coralligenous habitats represent an important $CO_2$ sink, playing a relevant role in the regulation of ocean acidification associated to global warming [25]. Moreover, the spectacular coralligenous formations distributed over 50 m depth (accessible to scuba diving activities) represent a great touristic attraction for their high esthetic value and are among the most preferred diving spots worldwide [33]. The main ecosystem goods and services provided by coralligenous habitats (*sensu* [34, 35]) are reported in **Table 1**.

## 2.2 CWC frameworks

There is not a general consensus about the use of the terms "reef" or "framework" to describe those marine bioconstructions structured by the so-called cold-water corals or white corals (subclass Hexacorallia, order Scleractinia), namely the colonial species *Madrepora oculata* Linnaeus, 1758 and *Lophelia pertusa* (Linnaeus, 1758) (recently renamed as *Desmophyllum pertusum*), as well as the solitary or pseudo-colonial coral *Desmophyllum dianthus* (Esper, 1794) (**Figure 1b**). These species can form extensive three-dimensional habitats in the Mediterranean Sea, from 180 to more than 1000 m depth [36], identified as biodiversity hot spots [37, 38]. Despite still far from being fully understood, the current known distribution of CWCs reveals some dozens of coral sites all over the Mediterranean Sea, form single occurrences to large CWC provinces [39]. These habitats provide a suitable hard ground for sessile species and act as shelter, feeding, spawning and nursery area for a variety of vagile species, representing an Essential Fish Habitat (EFH) for several commercial and non-commercial fish and invertebrate species [22, 40–49] (**Table 1**). For instance, it

| Category | Ecosystem goods and services | Habitat | | | |
|---|---|---|---|---|---|
| | | Coralligenous | CWC framework | Coral forest | Sea pen field |
| Supporting | Habitat forming | x | x | x | x |
| | Larval and gamete supply | x | x | x | x |
| | Breeding, spawning and nursery area | x | x | x | x |
| | Refuge and settling potential | x | x | x | x |
| | Primary production | x | | | |
| | Nutrient cycling | x | x | x | x |
| Provisioning | Food provision | x | x | x | x |
| | Active substances | x | x | x | x |
| | Genetic resources | x | x | x | x |
| Regulating | Climate regulation ($CO_2$ trapping) | x | x | | |
| | Biological control | x | x | x | x |
| | Disturbance regulation | x | x | | |
| | Bioremediation of waste | x | | | |
| | Erosion control and sediment retention | x | x | x | |
| Cultural | Recreation and tourism | x | | | |
| | Research and education | x | x | x | x |
| | Esthetic | x | | | |

**Table 1.**
*Ecosystem goods and services provided by the mesophotic and deep-sea vulnerable habitats of the Mediterranean Sea characterized by corals.*

has been recognized that the presence of CWC habitats benefits adjacent fisheries in the central Mediterranean [50]. Moreover, the massive amount of calcified colonies that constitute the bioconstruction represents an important $CO_2$ sink and, between the branched colonies, large quantities of sediment and larvae are also retained (**Table 1**). Furthermore, it is becoming well known that CWC frameworks are hot spots of global biogeochemical cycling [51–53].

Solitary scleractinians such as, among others, *Stenocyathus vermiformis* (Pourtalès, 1868), *Javania cailleti* (Duchassaing & Michelotti, 1864), *Anomocora fecunda* (Pourtalès, 1871) and *Caryophyllia* spp. can be present in CWC habitats, but they have not been reported so far with a relevant aggregative behavior and their role in the bioconstruction is often minimal [54].

The colonial yellow coral *Dendrophyllia cornigera* (Lamarck, 1816) can occur on flat or gently sloping hard bottoms, as well as on flat muddy bottoms without any consistent anchorage, from the mesophotic to the aphotic zone [39]. This CWC species can occasionally form coral habitats that are mostly known as *Dendrophyllia* beds rather than frameworks, because the density of the colonies does not reach high-enough values to give the appearance of a compact structure to the habitat [19]. On the contrary, the congeneric *Dendrophyllia ramea* (Linnaeus, 1758) has a shallower distribution (from 80 to more than 700 m depth, although more common within 200 m depth) and it is present both on hard and sedimentary bottoms, as well as within coralligenous formations [54, 55].

The hydrocoral *Errina aspera* (Linnaeus, 1767) can also be included in the CWCs *sensu lato*, because it is an habitat-former species of the mesophotic and the upper aphotic zones, where it can occasionally form monospecific stands showing high densities and being similar to CWC habitats [56].

## 2.3 Coral forests

Coral forests are marine habitats created by the aggregation of arborescent coral colonies belonging to one or few species of alcyonaceans and antipatharians [39]. These communities can develop both on hard and soft substrata, including detritic bottoms with small scattered substrata, such as small rocks, shells or coral rubble. Hard-bottom coral forests can be settled on both mesophotic rocky bottoms and deep coralligenous bioconstructions (**Figure 1a** and **c**), while in the deep sea, they can develop on rocky bottoms, hardgrounds or CWC frameworks [57–60] (**Figure 1d**). Antipatharians, also known as black corals (subclass Hexacorallia, order Antipatharia), form monospecific or multispecific forests on hard bottoms in both mesophotic and aphotic zones. In particular, *Antipathella subpinnata* (Ellis & Solander 1786) is much common on mesophotic bottoms (**Figure 1c**), *Antipathes dichotoma* Pallas, 1766 and *Parantipathes larix* (Esper, 1788) thrive from the mesophotic to the upper aphotic zones, while *Leiopathes glaberrima* (Esper, 1788) develops mostly on deep bottoms [20, 37, 39, 57, 60, 61].

Several species of alcyonaceans (subclass Octocorallia, order Alcyonacea) are present on hard bottoms, covering a wide bathymetric range (from 15 to more than 1000 m in depth, depending on the species). For instance, the gorgonians *Paramuricea clavata* (Risso, 1826) (**Figure 1a**), *Ellisella paraplexauroides* Stiasny, 1936 and *Eunicella* spp. are among the most common species that form large aggregations on mesophotic coralligenous habitats, as well as on other coherent substrata, often on the continental shelf [59, 62–65]. *Acanthogorgia hirsuta* Gray, 1857, *Callogorgia verticillata* (Pallas, 1766) (**Figure 1d**), *Paramuricea macrospina* (Koch, 1882), *Placogorgia coronata* Carpine & Grasshoff, 1975, *Viminella flagellum* (Johnson, 1863) and the precious red coral *Corallium rubrum* (Linnaeus, 1758) can act as habitat formers from mesophotic to deep bottoms [39, 57, 66–68].

On soft bottoms, the bamboo coral *Isidella elongata* (Esper, 1788) can form extensive populations on compact mud between 110 and 1600 m depth, on relatively flat or gently inclined seabed [69] (**Figure 1e**). This typology of substratum is also suitable for bottom trawling, which has a strong mechanic impact on *I. elongata* colonies as well as on other sessile species. For this reason, the Mediterranean populations of *I. elongata* are in strong decline and the species has been categorized as "critically endangered" by IUCN [16]. Fortunately, some populations of the species are currently surviving in places where trawling is not carried out [39, 69].

The presence of a coral forest enhances the three-dimensionality of the seabed, often representing a hot spot of biodiversity. In fact, some recent new records of deep-sea invertebrates in the Mediterranean Sea occurred within *I. elongata* populations [70–72]. Among the suite of ecosystem goods and services that these habitats provide (**Table 1**), they are known to act as refuge, feeding, spawning and nursery areas for many associated species, including species of commercial value [69, 73–75].

## 2.4 Sea pen fields

Sea pens (subclass Octocorallia, order Pennatulacea) live on soft bottoms, from shallow to deep waters, and they can form dense aggregations known as sea pen fields which increase the complexity of an otherwise flat and monotonous seabed [76, 77]. *Pennatula rubra* (Ellis, 1761) (**Figure 1f**), *Pteroeides spinosum* (Ellis, 1764) and *Veretillum cynomorium* (Pallas, 1766) are typical field-forming species of the continental shelf, being occasionally present form the euphotic zone but forming true fields only in the mesophotic [24, 78–80]. *Pennatula phosphorea* Linnaeus, 1758 and *Virgularia mirabilis* (Müller, 1776) can occur from mesophotic to deep seabed, often in mixed aggregation with other soft-bottom anthozoans, while *Funiculina quadrangularis* (Pallas, 1766) and *Kophobelemnon stelliferum* (Müller, 1776) represent deep-sea pennatulaceans able to form dense fields on the aphotic muddy bottoms [31, 81, 82].

Sea pen fields can act as refuge and nursery area for small-size specimens of many taxa, among which some crustaceans and fish, attracting the vagile fauna and enhancing the biodiversity of the area [24, 82–84] (**Table 1**).

## 3. Conservation status

Coral habitats are considered valuable and productive fishing areas due to the presence and abundance of species of commercial interest, whose capture using bottom-contact or bottom-tending fishing gears represents the major threats to the habitats [22, 85]. These gears can be towed (e.g., trawl nets and dredges) or fixed (e.g., longlines, gillnets, trammel nets, traps and pots; although static on the seabed, they may be pulled across the bottom for short distances during retrieval or storms).

Trawl nets, as well as bottom-contact longlines, gillnets and trammel nets are the most destructive fishing gears in the mesophotic and aphotic zones of the Mediterranean Sea. Trawling has a high mechanical impact on soft-bottom communities, while on hard bottoms, it affects the vulnerable filter- and suspension-feeder fauna both directly (e.g., sediment resuspension, destruction of benthos, dumping of processing waste) and indirectly (e.g., post-fishing mortality and long-term trawl-induced changes to the benthos) [85–87]. Bottom longlines, widely deployed all over the basin, can have a significant mechanical impact and their coral bycatch can be high, particularly during retrieval operations [42, 74]. Moreover, longlines can easily remain entangled in the rocky bottoms, thus becoming lost and damaging the benthic communities. Artisanal fishing practices, such as gillnets and trammel nets, are usually deployed on coastal areas in close proximity of cliffs and steep topographies; thus, the chance to catch corals and the possibility of being entangled on them are high. Mechanical injuries on benthic species are even higher when the entangled tools are abandoned on the seabed, altering habitats [85].

Trawling is forbidden in areas deeper than 1000 m depth all over the Mediterranean Sea. However, this conservation measure is not enough, considering that the majority of the coral habitats known so far is present within this depth limit [36, 39]. For this reason, the limitation of trawling up to 800 m depth would be more effective for the conservation of deep-sea coral habitats. During the last decades, the deep-sea habitats have been protected through the institution of Fishery Restricted Areas (FRAs), established by the General Fishery Commission for the Mediterranean and Black Sea (GFCM) with the aims of protecting VMEs and/or Essential Fish Habitats (EFHs). However, only one of the six existing FRAs of the Mediterranean Sea has been created to target the conservation of a CWC habitat, such as the *Lophelia* Reef off Santa Maria di Leuca (Italy, Ionian Sea) [21]. Two FRAs, namely the Strait of Sicily (northeast and northwest Malta) and the Gulf of Lion, include some CWC sites but they have been created to manage fishing stocks; so trawling is present although somehow regulated. The Jabuka/Pomo Pit FRA aims to protect EFHs and an unquantified sea pen field [88], while the Eratosthenes Seamount FRA targets the protection of peculiar geologic formations

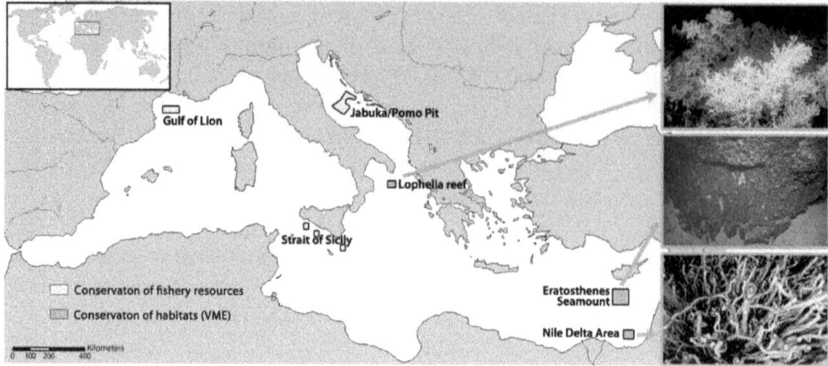


**Figure 2.**
*Distribution of the deep-sea fishery restricted areas in the Mediterranean Sea, three of them created for the management of fishery resources and three for the conservation of benthic habitats (vulnerable marine ecosystems, VMEs).*

(with only few specimens of solitary sleractinians recorded) [89] and the Nile Delta FRA is characterized by the presence of chemosynthetic fauna (**Figure 2**). Trawling and dredging are forbidden in the FRAs for the conservation of VMEs, while they are regulated in those for the management of EFHs. Bottom longlining can be allowed, often in a buffer zone and under authorization, depending on the regulamentation of the single FRA, while artisanal fishing practices are usually not performed in offshore areas as the existing FRAs. These FRAs result actually isolated, while a desirable network of FRAs is far to be created. This network should be established in the pathway of the Mediterranean water mass circulation in order to connect the different FRAs all over the basin by means of larval dispersal ([39] and references therein). For instance, along the southern Apulian margin, an almost continuous belt of CWC communities are crossed by the water masses that flow between the Southern Adriatic and Northern Ionian Seas [45, 58, 81, 90]. In this respect, the establishment of a FRA for the Bari Canyon has been proposed and the relative institution process is in progress [50].

The analogues of the Mediterranean FRAs are the Offshore Special Areas for Conservation in the North-Atlantic, such as the Darwin Mounds (northwest of Scotland), whose CWC ecosystem was discovered in 1998 and, from 2003, bottom-contact fisheries were banned in the area [91]. Also in the Sula Reef (Norway), one of the first reefs of *L. pertusa* discovered, trawling activities are banned [87].

Several actions have been undertaken also in the United States to address impacts of fishing on VMEs, such as banning of bottom trawls throughout the large area of the Western Pacific Fishery Management Council (3.9 million $km^2$) [92], as well as adoption of closed areas on Georges Bank in New England (Northwest Atlantic) [93]. Other solutions can include the shifting from gears with higher impacts to gears with lower impacts, as happened along the coast of California, where bycatch in the fishery of the California spot prawn *Pandalus platyceros* J.F. Brandt in von Middendorf, 1851 has been greatly reduced by shifting from bottom trawls to traps [94].

In the Caribbean, the Parque Nacional Natural Corales de Profundidad (Colombia) is the only deep-sea coral area actually under protection measures [95, 96].

Further extra-Mediterranean examples of effective conservation initiatives are those carried out by the Department of Fisheries and Oceans (DFO) in the North Atlantic which, during 2002, implemented its first fisheries closure to protect CWC populations in different areas of the Northeast Channel Coral Conservation Area (southwest of Nova Scotia, Canada) [97]. The management measures used in each area are different and the permitted activities vary from fishing regulation (though a limited number of licenses released and only for certain practices) to fishing closures and relative buffer areas, depending on each area features. In specific areas, such as the Northeast Channel Coral Conservation Area, a precautionary zone where fishing is not allowed has been identified below 500 m [97]. Management activities include logbooks, at-sea observers, vessel monitoring systems and surveillance overflights against illegal fishing practices. For instance, in 2004, DFO successfully prosecuted a longline fisher who did not have a fisheries observer onboard while fishing in the limited fisheries zone of the Northeast Channel [97]. Canada has undertaken a series of successful conservation actions also in the waters off British Columbia (Northeast Pacific) [98], with the concomitant adoption of four measures, such as: (1) identification of ecosystem-based spatial boundaries for the bottom trawl fleet to prevent further spatial expansion of fishery areas, exclusion of areas of historically high coral and sponge bycatch, as well as control that the area opened to bottom trawling did not disproportionately impact any single habitat type; (2) establishment of an habitat quota (i.e., a bycatch quota for habitat formers) through onboard observers; (3) development of an encounter protocol, such as a set of defined management steps to be applied when a fishing vessel encounters a significant amount of VME indicator taxa; and (4) implementation of the monitoring systems and formation of a scientific committee. Ten years before, the National Marine Fisheries Service in Alaska formulated a habitat protection alternative for the Aleutian Islands (North Pacific) based on an interdisciplinary cost-effective model for mitigating the adverse effects of fishing on deep sea developed by Oceana [99]. Working with the fishery operators, Oceana identified and delimited the actual fishing grounds where bottom-contact gears are deployed, avoiding the exploitation of further areas. This approach also uses observer data to identify VMEs, mostly as areas of high coral and sponge bycatch rates, to develop a comprehensive management policy that allows bottom trawling only in specific designated areas with current low habitat impacts and consistent fish harvest. All areas not specified as open and historically not exploited are closed to bottom trawling [99]. The Oceana's approach also includes coral and sponge bycatch limits in the trawled areas, as well as a plan for comprehensive seafloor research, mapping and monitoring [99]. Its enforcement is based on increased observer coverage, vessel monitoring systems and electronic logbooks, as well as its application for other fishing practices such as bottom longlining and artisanal fishery. The same approach has been adopted in the United States [100], freezing the existing bottom trawl footprint, regulating areas that have low fishing effort, closing sensitive areas such as coral gardens and seamounts and calling for further research and monitoring. According to the freezing-the-footprint approach, also the National Oceanic and Atmospheric Administration (NOAA) proposed a precautionary approach to manage adverse impacts of fishing on VMEs [101].

In the Mediterranean Sea, MPAs usually do not overlap with deep-sea VMEs, with the exception of the National Park of Calanques (France) and the Chella Bank (Seco de los Olivos, Spain); the latter declared also a Site of Community Importance (SCI) [102]. Mesophotic VMEs can be more or less accidentally included in MPAs, sometimes at the borders of the protected area and without a real awareness of their presence. On the contrary, a habitat-based maritime spatial planning of MPAs is needed to guarantee the strategic protection of mesophotic important sites, such as particular coralligenous formations, as well as forests of black corals and alcyonaceans. Although several MPAs have been designated where seascapes are particularly attractive for scuba divers and "beauty" has been the main reason for their proposal [103], including some spectacular coralligenous formations, in some areas, it is still present a limited overlap between the most visited diving sites and the presence of some forms of regulation, suggesting the presence of further areas worth of protection [33].

Generally speaking, most of the VMEs identified so far are not included in the existing FRAs and MPAs, as well as in others local form of protection. New conservation measures, which are comprehensive of the high diversity of habitats present all over the Mediterranean Sea, are difficult to be achieved. Recently, several attempts have been made to identify those strategic areas whose protection is of priority importance [6, 33, 39, 88], although they were biased by the targeted habitats, with different scenarios depending on whether the priority is given to deep-sea vs. coastal habitats, soft-bottom vs. hard-bottom communities, etc. For instance, bioconstructions such as coralligenous and CWC frameworks are often the result of centuries or even millennia of biological activities; their destruction can be almost irreversible, and for this reason, they require the greatest attention in any conservation measures [6]. Black coral forests can take long time to be formed and are highly sensitive to artisanal fishery, as well as to indirect effects from trawl fishing; thus, protected areas should be created to protect the last surviving forests of the Mediterranean Sea [104]. Soft-bottom habitats such as *I. elongata* gardens are the most sensitive to fishing pressures, particularly to trawling; they are critically endangered, representing a priority for conservation measures [39, 69]. In the same way, sea pen fields are sensitive to trawl fishing and are very important to be protected [88, 105]. All these priorities could be confounding for decision-makers, and usually it is not possible to give a hierarchy of priorities for conservation since all the habitats mentioned above are under urgent need for protection.

Despite the fragmented geopolitical scenario, the heterogeneity of fishing practices and the different fishery resources and local settings characterizing the Mediterranean, the abovementioned conservation approach adopted in Northeast and North Pacific could be taken into account also in this basin for a future comprehensive protection of VMEs from bottom-fishing activities, particularly trawling and longlining (**Figure 3**). The logbooks and Vessel Monitoring by satellite System

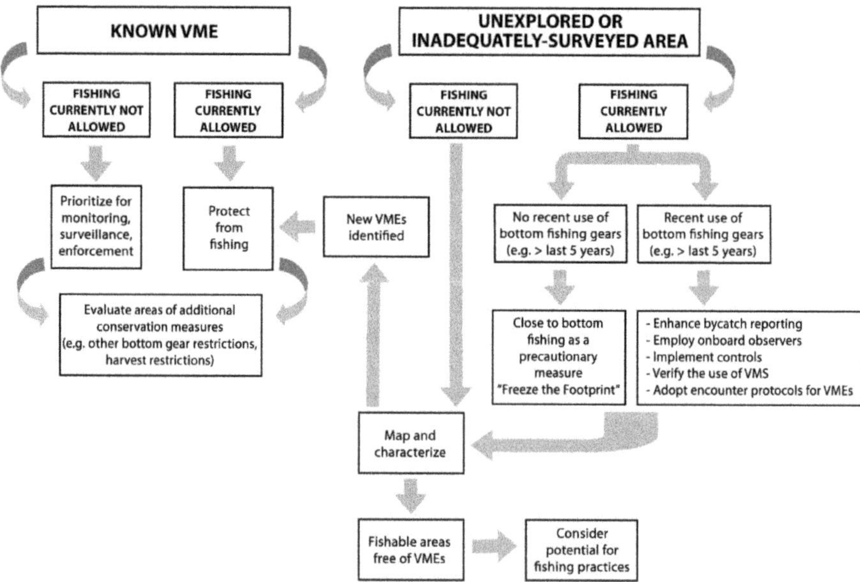


**Figure 3.**
*A scheme of a precautionary approach to manage bottom-tending fishing gears that could be applicable on Mediterranean vulnerable marine ecosystems. Modified from [101].*

(VMS), together with the closed season and closed areas, such as FRAs, are measures already adopted for the management of fishery resources on the northern side of the basin. All these measures could combine VME conservation and fisheries management objectives according to the Ecosystem Approach to Fisheries (EAF) [106].

In European waters, these measures are coupled with a large-scale monitoring of the fishing impact as part of the Marine Strategy Framework Directive [107]. The existing measures related to fisheries management could be implemented to include the assessment of VME incidental catches and to enforce controls for illegal fishing activities in restricted areas. However, the complex geopolitical setting of the basin, the distance of the fishing grounds from the coasts and the lack of proper controls do not favor its implementation. For instance, logbooks have been formally adopted, but the information reported is often incomplete and unreliable, for both commercial catches and discard. The use of observers onboard is still missing in commercial fishery, while it is limited to few research projects for a short time. Moreover, VMS data are not easily accessible and are scantly used for control, although they can reveal certain illegal fishing activities inside a FRA [21].

The regular use of observers onboard the fishing boats could help to collect a suite of reliable data about catches of commercial species and bycatch of vulnerable or protected species, as well as to avoid infringements and illegal fishery on VMEs or no-fishing areas. Observers would be also involved in the adoption and implementation of commercial encounter protocols for VMEs in order to limit impacts by fishing practices [92, 98]. These protocols can be set up after *ad hoc* studies that assess a maximum distance that a vessel should have to move after the catch (either shallower or deeper) of a specific quota of VME indicator taxa, which would vary depending on the species. Considering that many areas of the deep Mediterranean Sea are still scantly explored or unknown, the report of VME incidental catches could improve our understanding about the occurrence and distribution of these habitats on a basin scale, representing a needed effort for a comprehensive conservation of VMEs, including coral habitats (**Figure 3**).

Management measures, such as encounter protocols with associate thresholds and closure areas, can be developed and implemented according to the FAO's International Guidelines for the Management of Deep-sea Fisheries [108] in order to ensure VME protection from significant adverse fishing impacts. The use of onboard observers and the correct adoption of digital logbooks could be applicable to trawl fishing vessels and to most of the deep-sea benthic longlining vessels, which must be equipped with VMS and/or Automated Identification Systems (AIS) in correct working order. On the contrary, the management and control of the numerous and small artisanal fishing boats, which use gillnets and trammel nets in mesophotic habitats, could be done though the designation of landing points, obligations of notice of arrival in port and control of landings.

## 4. Conclusions

Although our understanding on the distribution and the main features of vulnerable coral habitats in the Mediterranean Sea is still incomplete, the information available is strong enough to support conservation strategies, based on what is currently known. Control enforcements, employment of onboard observers and implementation of a network of protected or fishing restricted areas are urgently needed measures to guarantee a proper conservation of these VMEs. A network of protected areas (mostly MPAs and FRAs) would satisfy both conservation of coral habitats and management of fishery resources according to the EAF. Conservation planning should take into account ecological issues and socioeconomic aspects related to the cost of preventing the use of these areas by humans. Public awareness, stakeholder involvement and a credible system of monitoring, control and surveillance will be fundamental to meet such conservation and management objectives. However, it is unlikely that all the Mediterranean benthic habitats will be comprehensively mapped in the near future. Thus, a precautionary approach for conservation is needed to ensure that, after the closure of a specific area to destructive fishing practices (at least trawling and longlining), fishing effort is properly managed and does not move into further areas that may be susceptible to damage (i.e., VMEs). This approach, combined with the identification of hot spots for conservation, would trigger an iterative process to develop effective management solutions to protect VMEs under a precautionary approach, combining habitat protection with maintaining commercial fishing opportunities.

## Acknowledgements

This chapter benefited from funding by the Italian Ministry of Education, University and Research (PON 2014-2020, AIM 1807508, Attività 1, Linea 1), the Ente Parco Nazionale del Gargano (Research agreement with CoNISMa N. 21/2018) and the National Geographic Society (Grant EC-176R-18).

## Conflict of interest

The authors declare no conflict of interest.

Author details

Giovanni Chimienti[1,2*], Francesco Mastrototaro[1,2] and Gianfranco D'Onghia[1,2]

1 Department of Biology, University of Bari Aldo Moro, Bari, Italy

2 CoNISMa, Roma, Italy

*Address all correspondence to: giovanni.chimienti@uniba.it

IntechOpen

## References

[1] Costello MJ, Coll M, Danovaro R, et al. A census of marine biodiversity knowledge, resources, and future challenges. PLoS One. 2010;**5**(8):e12110

[2] Danovaro R, Company JB, Corinaldesi C, et al. Deep-sea biodiversity in the Mediterranean Sea: The known, the unknown, and the knowable. PLoS One. 2010;**5**:e11832

[3] Abadie A, Pace M, Gobert S, Borg JA. Seascape ecology in *Posidonia oceanica* seagrass meadows: Linking structure and ecological processes for management. Ecological Indicators. 2018;**87**:1-13

[4] Ballesteros E, Sala E, Garrabou J, Zabala M. Community structure and frond size distribution of a deep water stand of *Cystoseira spinosa* (Phaeophyta) in the northwestern Mediterranean. European Journal of Phycology. 1998;**33**:121-128

[5] Basso D, Babbini L, Ramos-Esplá AA, Salomidi M. Mediterranean rhodolith beds. In: Riosmena-Rodríguez R, Nelson W, Aguirre J, editors. Rhodolith/ Maërl Beds: A Global Perspective. AG Cham, Switzerland: Springer International Publishing; 2018. pp. 281-298

[6] Ingrosso G, Abbiati M, Badalamenti F, et al. Mediterranean bioconstructions along the Italian coast. Advances in Marine Biology. 2018;**79**:61-136

[7] Rossi S, Bramanti L, Gori A, Orejas C, editors. Marine Animal Forests. The Ecology of Benthic Biodiversity Hotspots. AG Cham, Switzerland:Springer International Publishing; 2017. 1366p

[8] EEC Reg. 43 On the Conservation of Natural Habitats and of Wild Fauna and Flora (Habitat Directive). 1992.

Available from: http://eur-lex.europa. eu/legal-content/EN/TXT/?qid=1494 715607826&uri=CELEX:31992L0043 [Accessed: August 10, 2019]

[9] Gattuso JP, Gentili B, Duarte CM, et al. Light availability in the coastal ocean: Impact on the distribution of benthic photosynthetic organisms and their contribution to primary production. Biogeosciences. 2006;**3**:489-513

[10] Lesser MP, Slattery M, Leichter JJ. Ecology of mesophotic coral reefs. Journal of Experimental Marine Biology and Ecology. 2009;**375**:1-8

[11] Kahng SE, Garcia-Sais JR, Spalding HL, et al. Community ecology of mesophotic coral reef ecosystems. Coral Reefs. 2010;**29**:255-275

[12] FAO. International Guidelines for the Management of Deep-sea Fisheries in the High Seas. Rome: FAO; 2009. 73p

[13] GFCM. Criteria for the identification of sensitive habitats of relevance for the management of priority species. In: Meeting of the Sub-Committee on Marine Environment and Ecosystems (SCMEE). 30 November–3 December 2009; Malaga, Spain; 2009

[14] GFCM. Report of the First Meeting of the Working Group on Vulnerable Marine Ecosystems (WGVME). 3-5 April 2017; Malaga, Spain; 2017. 45p

[15] GFCM. Report of the Second Meeting of the Working Group on Vulnerable Marine Ecosystems (WGVME). 26-28 February 2018; Rome, Italy; 2018. 57p

[16] Otero MM, Numa C, Bo M, et al. Overview of the Conservation Status of Mediterranean Anthozoans. Málaga: IUCN; 2017. 73p

[17] Orejas C, Gori A, Lo Iacono C, et al. Cold-water corals in the cap de Creus canyon, northwestern Mediterranean: Spatial distribution, density and anthropogenic impact. Marine Ecology Progress Series. 2009;**397**:37-51

[18] Ramirez-Llodra E, Tyler PA, Baker MC, et al. Man and the last great wilderness: Human impact on the deep sea. PLoS One. 2011;**6**:e22588

[19] Bo M, Bava S, Canese S, et al. Fishing impact on deep Mediterranean rocky habitats as revealed by ROV investigation. Biological Conservation. 2014;**171**:167-176

[20] Bo M, Bavestrello G, Angiolillo M, et al. Persistence of pristine deep-sea coral gardens in the Mediterranean Sea (SW Sardinia). PLoS One. 2015;**10**(3):e0119393

[21] D'Onghia G, Calculli C, Capezzuto F, et al. Anthropogenic impact in the Santa Maria di Leuca cold-water coral province (Mediterranean Sea): Observations and conservation straits. Deep-Sea Research Part 2. 2017;**145**:87-101

[22] D'Onghia G. Cold-water corals as shelter, feeding and life-history critical habitats for fish species: Ecological interactions and fishing impact. In: Orejas C, Jiménez C, editors. Mediterranean Cold-Water Corals: Past, Present and Future. Chapter 30. Coral Reefs of the World 9. AG Cham, Switzerland: Springer International Publishing; 2019. pp. 335-356

[23] Orejas C, Jimenez C, editors. Mediterranean Cold-Water Corals: Past, Present and Future. Coral Reefs of the World 9. AG Cham, Switzerland: Springer International Publishing; 2019. 582p

[24] Chimienti G, Angeletti L, Rizzo L, et al. ROV vs trawling approaches in the study of benthic communities: The case of *Pennatula rubra* (Cnidaria:

Pennatulacea). Journal of the Marine Biological Association of the UK. 2018;**98**(8):1859-1869

[25] Ballesteros E. Mediterranean coralligenous assemblages: A synthesis of present knowledge. Oceanography and Marine Biology: An Annual Review. 2006;**44**:123-195

[26] Martin CS, Giannoulaki M, De Leo F, et al. Coralligenous and maërl habitats: Predictive modelling to identify their spatial distributions across the Mediterranean Sea. Scientific Reports. 2014;**4**:5073

[27] Marion AF. Esquisse d'une topographie zoologique du Golfe de Marseille. Annales du Musée d' Histoire Naturelle de Marseille. 1883;**1**(2):1-50

[28] Pruvot G. Sur les fonds sous-marins de la région de Banyuls et du cap de Creus. Comptes Rendus de l'Académie des. Sciences. 1894;**118**:203-206

[29] Pruvot G. Coup d'oeil sur la distribution générale des invertébrés dans la région de Banyuls (Golfe du Lion). Archives de Zoologie Expérimentale et Générale. 1895;**3**:629-658

[30] Pruvot G. Essai sur les fonds et la faune de la Manche Occidentale (côtes de Bretagne) comparées à ceux du Golfe de Lion. Archives de Zoologie Expérimentale et Générale. 1897;**5**:511-660

[31] Pérès JM, Picard J. Nouveau manuel de bionomie benthique de la mer Méditerranée. Recueil des Travaux de la Station Marine d'Endoume. 1964;**31**:1-137

[32] Cavanagh RD, Gibson C. Overview of the Conservation Status of Cartilaginous Fishes (Chondrichthyans) in the Mediterranean Sea. IUCN Species Survival Commission. Gland,

Switzerland and Malaga, Spain IUCN, Centre for Mediterranean Cooperation; 2007. 42p

[33] Chimienti G, Stithou M, Dalle Mura I, et al. An explorative assessment of the importance of Mediterranean Coralligenous habitat to local economy: The case of recreational diving. Journal of Environmental Accounting and Management. 2017;5(4):310-320

[34] Millennium Ecosystem Assessment (MA). Ecosystems and Human Well-Being: A Framework for Assessment. Washington DC: Island Press; 2003

[35] Liquete C, Piroddi C, Drakou EG, et al. Current status and future prospects for the assessment of marine and coastal ecosystem services: A systematic review. PLoS One. 2013;8(7):e67737

[36] Chimienti G, Bo M, Mastrototaro F. Know the distribution to assess the changes: Mediterranean cold-water coral bioconstructions. Rendiconti Lincei Scienze Fisiche e Naturali. 2018;29:583-588

[37] Mastrototaro F, D'Onghia G, Corriero G, et al. Biodiversity of the white coral ecosystem off cape Santa Maria di Leuca (Mediterranean Sea): An update. Deep-Sea Research Part 2. 2010;57:412-430

[38] Watling L, France SC, Pante E, Simpson A. Biology of deep-water octocorals. Advances in Marine Biology. 2011;60:41-123

[39] Chimienti G, Bo M, Taviani M, Mastrototaro F. Occurrence and biogeography of Mediterranean cold-water corals. In: Orejas C, Jiménez C, editors. Mediterranean Cold-Water Corals: Past, Present and Future. Chapter 19. Coral Reefs of the World 9. AG Cham, Switzerland: Springer International Publishing; 2019. pp. 213-243

[40] D'Onghia G, Maiorano P, Sion L, et al. Effects of deep-water coral banks on the abundance and size structure of the megafauna in the Mediterranean Sea. Deep-Sea Research Part 2. 2010;57:397-411

[41] D'Onghia G, Indennidate A, Giove A, et al. Distribution and behaviour of the deep-sea benthopelagic fauna observed using towed cameras in the Santa Maria di Leuca cold water coral province. Marine Ecology Progress Series. 2011;443:95-110

[42] D'Onghia G, Maiorano P, Carlucci, et al. Comparing deep-sea fish fauna between coral and non-coral "megahabitats" in the Santa Maria di Leuca cold-water coral province (Mediterranean Sea). PLoS One. 2012;7(9):e44509

[43] D'Onghia G, Capezzuto F, Cardone F, et al. Macro- and megafauna recorded in the submarine Bari canyon (southern Adriatic, Mediterranean Sea) using different tools. Mediterranean Marine Science. 2015;16(1):180-196

[44] D'Onghia G, Capezzuto F, Carluccio A, et al. Exploring composition and behaviour of fish fauna by *in situ* observations in the Bari canyon (southern Adriatic Sea, Central Mediterranean). Marine Ecology. 2015;36:541-556

[45] D'Onghia G, Calculli C, Capezzuto F, et al. New records of cold-water coral sites and fish fauna characterization of a potential network existing in the Mediterranean Sea. Marine Ecology. 2016;37(6):1398-1422

[46] Capezzuto F, Ancona F, Carlucci R, et al. Cold-water coral communities in the Central Mediterranean: Aspects on megafauna diversity, fishery resources and conservation perspectives. Rendiconti Lincei Scienze Fisiche e Naturali. 2018;29(3):589-597

[47] Capezzuto F, Sion L, Ancona F, et al. Cold-water coral habitats and canyons as essential fish habitats in the southern Adriatic and northern Ionian Sea (Central Mediterranean). Ecological Questions. 2018;**29**(2):9-23

[48] Capezzuto F, Calculli C, Carlucci R, et al. Revealing the coral habitat effect on the bentho-pelagic fauna diversity in the Santa Maria di Leuca cold-water coral province using different devices and Bayesian hierarchical modelling. Aquatic Conservation. 2019; **29**(10):1608-1622

[49] Sion L, Calculli C, Capezzuto F, et al. Does the Bari canyon (Central Mediterranean) influence the fish distribution and abundance? Progress in Oceanography. 2019;**170**:81-92

[50] D'Onghia G, Sion L, Capezzuto F. Cold-water coral habitats benefit adjacent fisheries along the Apulian margin (Central Mediterranean). Fisheries Research. 2019;**213**:172-179

[51] Van Oevelen D, Duineveld GCA, Lavaleye MSS, et al. The cold-water coral community as hotspot of carbon cycling on continental margins: A food web analysis from Rockall Bank (Northeast Atlantic). Limnology and Oceanography. 2009;**54**:1829-1844

[52] Cathalot C, Van Oevelen D, Cox TJS, et al. Cold-water coral reefs and adjacent sponge grounds: Hotspots of benthic respiration and organic carbon cycling in the deep sea. Frontiers in Marine Science. 2015;**2**(37):1-12

[53] Rovelli L, Attard K, Bryant LD, et al. Benthic $O_2$ uptake of two cold-water coral communities estimated with thenon-invasive eddy correlation technique. Marine Ecology Progress Series. 2015;**525**:97-104

[54] Zibrowius H. Les scléractiniaires de la Méditerranée et de l'Atlantique nord-oriental. Mémoires de l'Institute Océanographique, Monaco. 1980;**11**:1-284

[55] Orejas C, Gori A, Jiménez C, et al. Occurrence and distribution of the coral *Dendrophyllia ramea* in Cyprus insular shelf: Environmental setting and anthropogenic impacts. Deep-Sea Research Part 2. 2019;**164**:190-205

[56] Salvati E, Angiolillo M, Bo M, et al. The population of *Errina aspera* (Hydrozoa: Stylasteridae) of the Messina Strait (Mediterranean Sea). Journal of the Marine Biological Association of the UK. 2010;**90**(7):1331-1336

[57] Bo M, Canese S, Spaggiari C, et al. Deep coral oases in the South Tyrrhenian Sea. PLoS One. 2012;**7**(11):e49870

[58] Angeletti L, Taviani M, Canese S, et al. New deep-water cnidarian sites in the southern Adriatic Sea. Mediterranean Marine Science. 2014;**15**(2):225-238

[59] Grinyó J, Gori A, Ambroso S, et al. Diversity, distribution and population size structure of deep Mediterranean gorgonian assemblages (Menorca Channel, Western Mediterranean Sea). Progress in Oceanography. 2016;**145**:42-56

[60] Cau A, Follesa MC, Moccia D, et al. *Leiopathes glaberrima* millennial forest from SW Sardinia as nursery ground for the small spotted catshark *Scyliorhinus canicula*. Aquatic Conservation. 2017;**27**:731-735

[61] Bo M, Bavestrello G, Canese S. Coral assemblage off the Calabrian coast (South Italy) with new observations on living colonies of *Antipathes dichotoma*. Italina Journal of Zoology. 2011;**78**(2):231-242

[62] Cerrano C, Danovaro R, Gambi C, et al. Gold coral (*Savalia savaglia*) and

gorgonian forests enhance benthic biodiversity and ecosystem functioning in the mesophotic zone. Biodiversity and Conservation. 2010;**19**:153-167

[63] Gori A, Rossi S, Berganzo E, et al. Spatial distribution patterns of the gorgonians *Eunicella singularis*, *Paramuricea clavata*, and *Leptogorgia sarmentosa* (cape of Creus, northwestern Mediterranean Sea). Marine Biology. 2011;**158**(1):143-158

[64] Gori A, Rossi S, Linares C, et al. Size and spatial structure in deep versus shallow populations of the Mediterranean gorgonian *Eunicella singularis* (cap de Creus, northwestern Mediterranean Sea). Marine Biology. 2011;**158**(8):1721-1732

[65] Gori A, Bramanti L, López-González P, et al. Characterization of the zooxanthellate and azooxanthellate morphotypes of the Mediterranean gorgonian *Eunicella singularis*. Marine Biology. 2012;**159**(7):1485-1496

[66] Bavestrello G, Bo M, Bertolino M, et al. Long-term structure and dynamics of the red coral community in the Portofino MPA. Marine Ecology. 2014;**36**(4):1354-1363

[67] Enrichetti F, Bavestrello G, Coppari M, et al. *Placogorgia coronata* first documented record in Italian waters: Use of trawl bycatch to unveil vulnerable deep-sea ecosystems. Aquatic Conservation. 2018;**28**:1123-1138

[68] De la Torriente A, Aguilar R, Serrano A, et al. Sur de Almería-Seco de los Olivos. Proyecto LIFE + INDEMARES. Fundación Biodiversidad del Ministerio de Agricultura, Alimentación y Medio Ambiente Madrid, Spain; 2014. 102p

[69] Mastrototaro F, Chimienti G, Acosta J, et al. *Isidella elongata* (Cnidaria: Alcyonacea) *facies* in the western Mediterranean Sea: Visual surveys and descriptions of its ecological role. European Zoological Journal. 2017;**84**(1):209-225

[70] Mastrototaro F, Chimienti G, Capezzuto F, et al. First record of *Protoptilum carpenteri* (Cnidaria: Octocorallia: Pennatulacea) in the Mediterranean Sea. The Italian Journal of Zoology. 2015;**82**(1):61-68

[71] Mastrototaro F, Chimienti G, Montesanto F, et al. Finding of the macrophagous deep-sea ascidian *Dicopia antirrhinum* Monniot, 1972 (Chordata: Tunicata) in the Tyrrhenian Sea and updating of its distribution. The European Zoological Journal. 2019;**86**(1):181-188

[72] Chimienti G, Aguilar R, Gebruk A, Mastrototaro F. Distribution and swimming ability of the deep-sea holothuroid *Penilpidia ludwigi* (Holothuroidea: Elasipodida: Elpidiidae). Marine Biodiversity. 2019:12. DOI: 10.1007/s12526-019-00973-9

[73] Fabri MC, Pedela L, Beuck L, et al. Megafauna of vulnerable marine ecosystems in French Mediterranean submarine canyons: Spatial distribution and anthropogenic impacts. Deep-Sea Research Part 2. 2014;**104**:184-207

[74] Mytilineou C, Smith CJ, Anastasopoulou A, et al. New cold-water coral occurrences in the eastern Ionian Sea: results from experimental long line fishing. Deep-Sea Research Part 2. 2014;**99**:146-157

[75] Mastrototaro F, Aguilar R, Chimienti G, et al. The rediscovery of *Rosalinda incrustans* (Cnidaria: Hydrozoa) in the Mediterranean Sea. The Italian Journal of Zoology. 2016;**83**(2):244-247

[76] Williams GC. The global diversity of sea pens (Cnidaria: Octocorallia: Pennatulacea). PLoS One. 2011;**6**:e22747

[77] Chimienti G, Angeletti L, Mastrototaro F. Withdrawal behaviour of the red sea pen *Pennatula rubra* (Cnidaria: Pennatulacea). European Zoological Journal. 2018;**85**(1):64-70

[78] Porporato EMD, Mangano MC, De Domenico F, et al. First observation of *Pteroeides spinosum* (Anthozoa: Octocorallia) fields in a Sicilian coastal zone (Central Mediterranean Sea). Marine Biodiversity. 2014;**44**:589-592

[79] Chimienti G, Tursi A, Mastrototaro F. Biometric relationships in the red sea pen *Pennatula rubra* (Cnidaria: Pennatulacea). In: Proceedings of the IEEE International Workshop on Metrology for the Sea; Learning to Measure Sea Health Parameters (MetroSea); 8-10 October 2018; Bari, Italy. 2018. pp. 212-216

[80] Chimienti G, Di Nisio A, Lanzolla AML, et al. Towards non-invasive methods to assess population structure and biomass in vulnerable sea pen fields. Sensors. 2019;**19**:2255

[81] Freiwald A, Beuck L, Rüggeberg A, et al. The white coral community in the Central Mediterranean Sea revealed by ROV surveys. Oceanography. 2009;**22**(1):58-74

[82] Mastrototaro F, Maiorano P, Vertino A, et al. A *facies* of *Kophobelemnon* (Cnidaria, Octocorallia) from Santa Maria di Leuca coral province (Mediterranean Sea). Marine Ecology. 2013;**34**:313-320

[83] Pardo E, Aguilar R, García S, et al. Documentación de arrecifes de corales de agua fría en el Mediterráneo occidental (Mar de Alborán). Chronica Naturae. 2011;**1**:20-34

[84] Baillon S, Hamel JF, Wareham VE, et al. Deep cold-water corals as nurseries for fish larvae. Frontiers in Ecology and the Environment. 2012;**10**(7):351-356

[85] Hinz H. Impact of bottom fishing on animal forests: Science, conservation, and fisheries management. In: Rossi S et al., editors. Marine Animal Forests: The Ecology of Benthic Biodiversity Hotspots. Switzerland: Springer; 2017. pp. 1041-1059

[86] Jones JB. Environmental impact of trawling on the seabed: A review. New Zealand Journal of Marine and Freshwater Research. 1992;**26**:59-67

[87] Fosså JH, Mortensen PB, Furevik DM. The deep-water coral *Lophelia pertusa* in Norwegian waters: Distribution and fishery impacts. Hydrobiologia. 2002;**471**:1-12

[88] Bastari A, Pica D, Ferretti F, et al. Sea pens in the Mediterranean Sea: Habitat suitability and opportunities for ecosystem recovery. ICES Journal of Marine Science. 2018;**75**(6):2289-2291

[89] Galil B, Zibrowius H. First benthos samples from Eratosthenes seamount, eastern Mediterranean. Marine Biodiversity. 1998;**28**(4):111-121

[90] Taviani M, Vertino A, López Correa M, et al. Pleistocene to recent scleractinian deep-water corals and coral facies in the eastern Mediterranean. Facies. 2011;**57**(4):579-603

[91] Huvenne VAI, Bett BJ, Masson DG, et al. Effectiveness of a deep-sea cold-water coral Marine protected area, following eight years of fisheries closure. Biological Conservation. 2016;**200**:60-69

[92] Code of Federal Regulations. Fisheries off west coast state and in the Western Pacific: Gear restrictions. Code of Federal Regulation. 2002;**50**:461-556

[93] Collie JS, Escanero GA, Valentine PC. Effects of bottom fishing

on the benthic megafauna of Georges Bank. Marine Ecology Progress Series. 1997;**35**:159-172

[94] Reilly PN, Geibel J. Results of California Department of Fish and Game Spot Prawn Trawl and Trap Fisheries Bycatch Observer Program, 2000-2001. Monterey, California, Belmont: California Department of Fish Game; 2002. p. 88

[95] Reyes J, Santodomingo N, Gracia A, et al. Southern Caribbean azooxanthellate coral communities off Colombia. In: Freiwald A, Roberts JM, editors. Cold-Water Corals and Ecosystems. Berlin Heidelberg: Springer-Verlag; 2005. pp. 309-330

[96] Urriago J, Santodomingo N, Reyes J. Formaciones coralinas de profundidad: criterios biológicos para la conformación de áreas marinas protegidas del margen continental (100-300 m) en el Caribe colombiano. Boletín de Investigaciones Marinas y Costeras. 2011;**40**(1):89-113

[97] Breeze H, Fenton DG. Designing management measures to protect cold-water corals off Nova Scotia, Canada. In: George RY, Cairns SD, editors. Conservation and Adaptive Management of Seamount and Deep-Sea Coral Ecosystems. Rosenstiel School of Marine and Atmospheric Science: University of Miami; 2007. pp. 123-133

[98] Wallace S, Turris B, Driscoll J, et al. Canada's pacific groundfish trawl habitat agreement: A global first in an ecosystem approach to bottom trawl impacts. Marine Policy. 2015;**60**:240-248

[99] Shester G, Ayers J. A cost effective approach to protecting deep-sea coral and sponge ecosystems with an application to Alaska's Aleutian Islands region. In: Freiwald A,

Roberts JM, editors. Cold-Water Corals and Ecosystems. Berlin Heidelberg: Springer-Verlag; 2005. pp. 1151-1169

[100] Shester G, Warrenchuk JUS. Pacific coast experiences in achieving deep-sea coral conservation and marine habitat protection. Bulletin of Marine Science. 2007;**81**(1):169-184

[101] NOAA. NOAA Strategic Plan for Deep-Sea Coral and Sponge Ecosystems: Research, Management, and International Cooperation. Silver Spring, MD: NOAA Coral Reef Conservation Program. NOAA Technical Memorandum CRCP; 2010;**11**:67

[102] Otero MM, Marin P. Conservation of cold-water corals in the Mediterranean: Current status and future prospects for improvement. In: Orejas C, Jiménez C, editors. Mediterranean Cold-Water Corals: Past, Present and Future. Chapter 46. Coral Reefs of the World 9. AG Cham, Switzerland: Springer International Publishing; 2019. pp. 535-545

[103] Boero F, Foglini F, Fraschetti S, et al. CoCoNET: Towards coast to coast networks of marine protected areas (from the shore to the high and deep sea), coupled with sea-based wind energy potential. Scientific Research and Information Technology. 2016;**6**:1-95

[104] Bo M, Tazioli S, Spanò N, Bavestrello G. *Antipathella subpinnata* (Antipatharia, Myriopathidae) in Italian seas. The Italian Journal of Zoology. 2008;**75**:185-195

[105] Kenchington E, Murillo FJ, Cogswell A, Lirette C. Development of encounter protocols and assessment of significant adverse impact by bottom trawling for sponge grounds and sea pen fields in the NAFO regulatory area. NAFO Scientific Council Report. 2011;**11**(75):53

[106] Garcia SM, Zerbi A, Aliaume C,
et al. The ecosystem approach
to fisheries. Issues, terminology,
principles, institutional foundations,
implementation and outlook. FAO
Fisheries Technical Paper. 2003;**443**:71

[107] EC Reg. 56 Establishing a
Framework for Community Action
in the Field of Marine Environmental
Policy (Marine Strategy Framework
Directive). 2008. Available from: http://
eur-lex.europa.eu/legal-content/
EN/TXT/?uri=CELEX:32008L0056
[Accessed: August 10, 2019]

[108] Armstrong CW, Foley NS, Kahui V,
et al. Cold water coral reef management
from an ecosystem service perspective.
Marine Policy. 2014;**50**:126-134

# Chapter 7

# Mitochondrial Group I Introns in Hexacorals Are Regulatory Genetic Elements

*Steinar Daae Johansen and Åse Emblem*

## Abstract

Hexacoral mitochondrial genomes are highly economically organized and vertebrate-like in size, structure, and gene content. A hallmark, however, is the presence of group I introns interrupting essential oxidative phosphorylation (OxPhos) genes. Two genes, encoding NADH dehydrogenase subunit 5 (ND5) and cytochrome c oxidase subunit I (COI), are interrupted with introns. The ND5 intron, located at position 717, is obligatory in all hexacoral specimens investigated. The ND5-717 intron is a giant-sized intron that carries several canonical OxPhos genes. Different modes of splicing appear to apply for the ND5-717 intron, including conventional *cis*-splicing, backsplicing, and *trans*-splicing. Three distinct versions of hexacoral COI introns are noted at genic positions 884, 867, and 720. The COI introns are of the mobile-type, carrying homing endonuclease genes (HEGs). Some COI-884 intron HEGs are highly expressed as in-frame COI exon fusions, while the expression of COI-867 intron HEGs appear repressed. We discuss biological roles of hexacoral mitochondrial ND5 and COI introns and suggest that the ND5-717 intron has gained new regulatory functions beyond self-splicing.

**Keywords:** backsplicing, colonial anemone, mitochondrial genome, mtDNA, mushroom corals, sea anemone, stony corals

## 1. Introduction

Hexacorallia (hexacorals) represents an ecological important subclass of Anthozoa with about 4300 extant nematocyst-bearing species [1]. Well-known hexacoral orders include Actiniaria (sea anemones), Zoantharia (colonial anemones), Scleractinia (stony corals), Corallimorpharia (mushroom corals), and Antipatharia (black corals). Ceriantharia (tube anemones) was previously considered to be a hexacoral order, but recent studies suggest tube anemones to represent a distinct subclass of Anthozoa [2].

Hexacorals have a global marine distribution pattern typically recognized in tropical seas at shallow waters living in close relationships with endosymbiotic photosynthetic alga. However, coral reefs and sea anemones in deep offshore waters have more recently been investigated [3–6]. These cold-water hexacorals occur in low temperatures at high latitudes or great depths. Among the approximately 1500 stony coral species known, 50% are located in cold-water habitats [7, 8]. A common feature among cold-water deep-sea hexacorals is that they are non-endosymbiotic in respect to the photosynthetic alga.

Mitochondria are essential organelles of animal cells, involved in processes like cell metabolism, cell signaling, and cell death [9, 10]. Hexacorals, like all other animals, contain mitochondrial genomes (mtDNAs) encoding a subset (approximately 1%) of the gene products involved in mitochondrial structure and function [11]. Complete mtDNA sequences have been determined from approximately 200 hexacoral specimens representing 133 species and 51 families from sea anemones, colonial anemones, stony corals, mushroom corals, and black corals (**Appendix Table 1**). In general, hexacoral mitochondrial genomes are vertebrate-like in size (17–22 kb), structure, and coding capacity (**Figure 1A**). The circular and economically organized mtDNA encodes the same set of 2 ribosomal RNAs and 13 hydrophobic proteins involved in the oxidative phosphorylation (OxPhos) system [11]. However, noncanonical and optional mitochondrial genes may occur in some hexacoral species [11–16]. More unusual features, however, are the highly reduced tRNA gene repertoire (only 1–2 tRNA genes) and the presence of complex group I introns [11, 17–20].

Group I introns are intervening sequences interrupting functional genes in eukaryotic (mitochondrial, chloroplast, nuclear, viral) and prokaryotic (eubacterial, archaeal, phage) genomes [21]. Like other mobile genetic elements, horizontal transfer of a group I intron can affect the host by altering the function of surrounding genes, potentially interrupting vital processes but also creating diversity and beneficial alterations. Mitochondrial group I introns in metazoans are rare and restricted to some orders within the basal phyla of Placozoa, Porifera, and Cnidaria [11, 22]. Unlike spliceosomal introns, which are abundant in the nuclear genome of eukaryotes, group I introns encode catalytic RNAs (ribozymes) with the unique

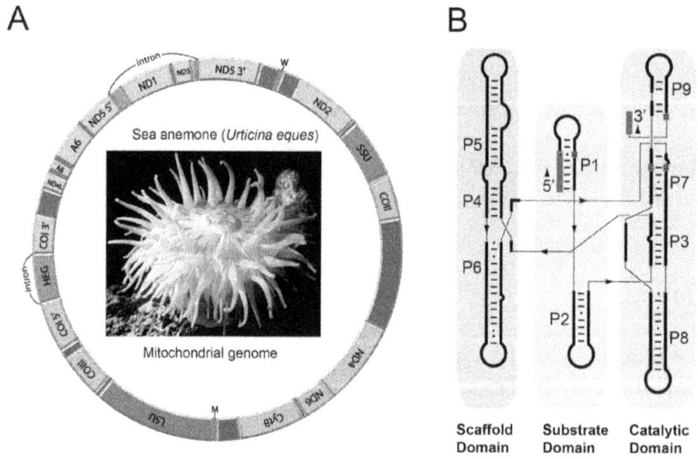


**Figure 1.**

*Mitochondrial genome and group I intron. (A) Circular map presenting gene content and organization of the sea anemone* Urticina eques *mtDNA. The mitochondrial genome harbors 14 protein coding genes, 2 rRNA genes, and 2 tRNA genes. All genes are encoded by the same DNA strand. The tRNA genes M and W (tRNA$^{fMet}$ and tRNA$^{Trp}$) are indicated by the standard one-letter symbols for amino acids; SSU and LSU, mitochondrial small- and large-subunit rRNA genes; ND1–6, NADH dehydrogenase subunit 1–6 genes; COI-III, cytochrome c oxidase subunit I–III genes; Cyt b, cytochrome b gene; ATP6 and 8, ATPase subunit 6 and 8 genes; and HEG, homing endonuclease gene. The ND5-717 and CO-884 introns are indicated. Photo: SD Johansen. (B) A general diagram of group I ribozyme secondary and tertiary structure, according to the representation by [23]. The nine conserved secondary structure paired segments of the catalytic core (P1–P9) are shown, and the three tertiary domains (scaffold, substrate, catalytic) are indicated by blue, yellow, and green boxes, respectively. Essential nucleotide positions in P1 (U, G), P7 (G, C), and P9 (G) are indicated in red. 5', upstream exon sequence; 3', downstream exon sequence.*

ability to self-splice as naked RNA. These introns sometimes even code for homing endonucleases, giving additional mobility to the ribozymes. The intron RNA processing reaction is catalyzed by the ribozyme, which folds into at least nine conserved paired segments (P1–P9), further organized into hallmark helical stacks named the catalytic domain, the substrate domain, and the scaffold domain (**Figure 1B**) [23–25]. Group I intron sequences are removed from precursor transcripts in a guanosine-dependent two-step transesterification reaction, leading to exon ligation and intron excision [21].

This chapter reviews recent developments in the characterization of hexacoral mitochondrial genomes with a focus on gene organization and rearrangements, complex obligatory group I introns in the NADH dehydrogenase subunit 5 (ND5) gene, and mobile-type group I introns in the cytochrome c oxidase subunit I (COI) gene.

## 2. Mitochondrial gene organization and expression in hexacorals

Five common features in the gene organization can be drawn from the 200 available mitochondrial genome sequences representing all five hexacoral orders (**Appendix Table 1**). (1) The 13 annotated OxPhos genes encode the same set of proteins as in vertebrate mtDNA [26], representing Complex I (ND1, 2, 3, 4, 4 L, 6, and 6), Complex III (CytB), Complex IV (COI, II, and III), and Complex V (ATPases 6 and 8). The additional approximately 70 OxPhos proteins are nuclear encoded [27]. (2) All canonical mitochondrial genes (OxPhos genes, rRNA, and tRNA genes) are encoded by the same DNA strand. (3) The tRNA gene repertoire is highly reduced, corresponding to tRNA$^{fMet}$ and tRNA$^{Trp}$ in sea anemones, stony corals, mushroom corals, and black corals, and only tRNA$^{fMet}$ in colonial anemones [14, 28, 29]. This indicates extensive tRNA import into mitochondria [20]. (4) The ND5 gene is split into two exons at nucleotide position 717 (human ND5 gene numbering [19]) by a group I intron found in all hexacorals studied so far (see Section 3 below). (5) The mitochondrial gene synteny appears highly conserved within, but not between, different hexacoral orders.

### 2.1 Order-specific gene organization

Each hexacoral order harbors a closely related primary mitochondrial gene organization (**Figure 2A**). This is an interesting notion since the orders have been separated from each other for 100 million years or more [28]. Stony corals and mushroom corals share some mtDNA synteny [28, 30], and similarly, some segments of synteny appear conserved between sea anemones, colonial anemones, and black corals [30]. The only mitochondrial gene synteny common to all species in all five orders is the upstream proximity of the tRNA$^{fMet}$ gene to the large-subunit (LSU) rRNA gene (**Figure 2**). This suggests co-expression similar to that of tRNA$^{Val}$ and LSU rRNA genes in vertebrate mitochondria [26]. Recent studies in human and rat conclude that the mitochondrial encoded tRNA$^{Val}$ has replaced the 5S rRNA and become an integrated component as a structural rRNA of the mitochondrial ribosome [31]. Thus, tRNA$^{fMet}$ is considered as an interesting candidate for a similar dual function in hexacorals.

Deviations from the primary order arrangements have been reported in some sea anemones, stony corals, and mushroom corals and apparently confined to non-endosymbiotic deep-water species. Among the stony corals (**Figure 2B**), *Madrepora* has a rearrangement in the COII and COIII gene order, and *Lophelia* and *Solenosmilia* have a more dramatic rearrangement involving three genes (CytB, ND2, and ND6) [19, 32]. The latter example involves a dramatic shift in the size of

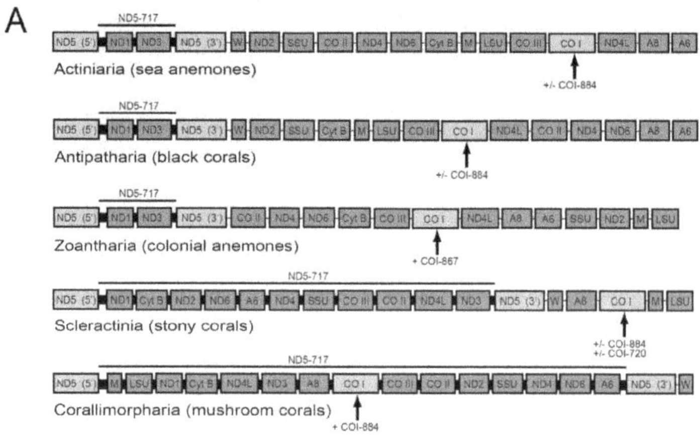

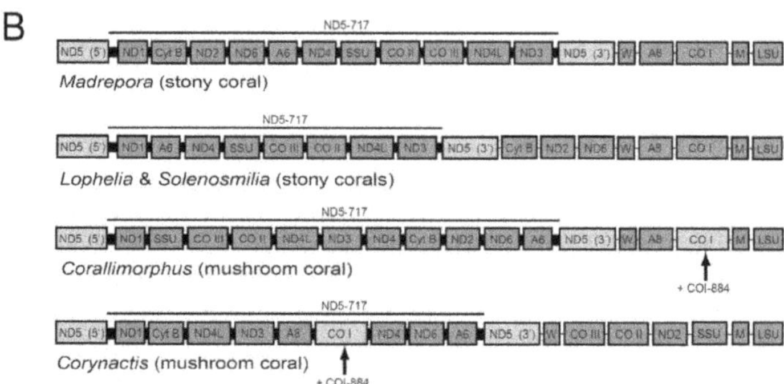

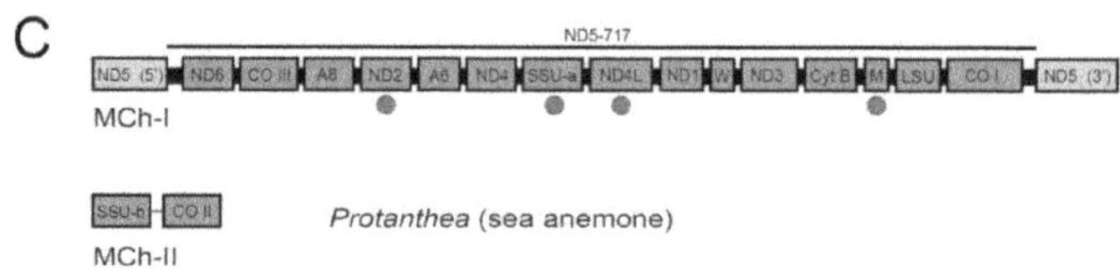

**Figure 2.**
*Gene organization of hexacoral mitochondrial genomes. Linear presentations of circular maps. Intron-containing OxPhos genes (yellow); intron-lacking OxPhos genes (blue); structural RNA genes (red). The obligatory ND5-717 introns are indicated by black lines, and the optional COI introns by arrows. (A) Primary arrangement in the five hexacoral orders Actiniaria, Antipatharia, Zoantharia, Scleractinia, and Corallimorpharia. (B) Deviations from the primary arrangement seen in the deep-water species* Madrepora oculata, Lophelia pertusa, Solenosmilia variabilis, Corallimorphus profundus, *and* Corynactis californica. *(C) Deviation from the sea anemone primary arrangement seen in the deep-water* Protanthea simplex. *MCh-I and MCh-II, mitochondrial chromosome I and II. SSU-a and SSU-b, two alleles of the small subunit ribosomal RNA gene. Genes located on the opposite strand in MCh-I are indicated by red dots.*

the ND5-717 intron from approximately 10 kb (primary arrangement) to 6 kb (see Section 3.1 about transfers of OxPhos genes into the intron). Two different deviations were noted in the mushroom corals *Corallimorphus* and *Corynactis* [30]. These rearrangements appear complex and involve a drastic size reduction of the ND5-717 intron from approximately 18 kb (primary arrangement) to 12 kb and 10 kb, respectively (**Figure 2B**).

The most dramatic mitochondrial genome rearrangement is seen in the deep-water sea anemone *Protanthea* [16]. Here, the 21 kb mtDNA is arranged along two circular mitochondrial chromosomes, MCh-I and MCh-II (**Figure 2C**). The mitochondrial gene order is heavily scrambled compared to the primary sea anemone arrangement. Different from all other hexacorals, genes at MCh-I are coded on both DNA strands. The ND5-717 intron size was increased from approximately 2 kb (primary sea anemone arrangement) to 15 kb in *Protanthea* (**Appendix Table 1**). Interestingly, the smaller MCh-II encodes the mitochondrial COII and one allele of the small subunit (SSU) rRNA. Phylogenetic analysis indicates that MCh-II is horizontally transferred into *Protanthea* from a distantly related sea anemone [16]. Not all deep-water hexacorals have mtDNA rearrangements. The *Relicanthus* sea anemone, sampled at a depth of 2500 m, [4] harbors the primary arrangement [33]. Similarly, *Bolocera* specimen samples at 40 m (Atlantic Ocean) [12] and at 1100 m (Pacific Ocean) [5] contain the same primary sea anemone arrangement.

## 2.2 Mitochondrial RNA in hexacorals

Mitochondrial RNAs have been investigated in a few hexacoral species representing sea anemones, colonial anemones, and mushroom corals [6, 12, 14–16]. RNAseq data were obtained from 454 pyrosequencing and Ion Torrent PGM sequencing. Several general features are noted: (1) ribosomal RNA constituted more than 90% of the reads and is found to be at least 10–20 times more abundant than most OxPhos gene transcripts; (2) all the conventional genes were transcribed, and the Complex IV OxPhos genes appeared most expressed; (3) group I introns were perfectly spliced out from ND5 and COI mRNA precursors; (4) COI-884 intron splicing appeared more efficient than that of the ND5-717 intron, suggesting intron retention of ND5 mRNA [16]; and (5) noncanonical mitochondrial genes, such as the intron-encoded HEG and non-annotated open reading frames (ORFs), were clearly expressed. One of these ORFs, corresponding to a 306-amino-acid unknown protein in the mushroom coral *Amplexidiscus*, was highly expressed and located at the opposite strand compared to canonical OxPhos genes [16].

## 3. An obligatory group I intron in the ND5 gene

All hexacoral mitochondrial genomes harbor ND5-717 introns (**Appendix Table 1**), making this group I intron an obligatory feature. Evolutionary analyses of ND5-717 introns have previously been performed and show a strict vertical inheritance pattern and a fungal origin [19]. Homologous group I introns at the ND5 insertion site 717 are frequently noted in the fungi Ascomycota, Basidiomycota, and Zygomycota [34], which include mobile-type versions with HEGs [35, 36]. This supports an ancient transfer with a subsequent progression into an obligatory strict vertical inherited intron. Interestingly, HEG-containing ND5-717 was also reported in the mitochondrial genome of choanoflagellates, species considered as the animal ancestors [37].

### 3.1 The ND5-717 intron is a giant group I intron

Phylogenetic analysis supports the early version of hexacoral ND5-717 introns to harbor two OxPhos genes (ND1 and ND3) in P8 [19]. This ancient organization is represented by sea anemones, colonial anemones, and black corals (**Figure 2A**). Insertions of ORFs into loop regions are a common feature in group I introns, and engulfing these compulsory genes might be a strategy for the intron in becoming essential to the host genome. RNA secondary structure folding of the ND5-717 ribozyme reveals that the catalytically important ωG (last nucleotide of the intron) is replaced by ωA (**Figure 3**). This replacement is likely to have a dramatic effect on intron biology, leading to host-factor dependent splicing and inhibition of 3′ hydrolysis-dependent intron RNA circularization [38].

In some hexacoral orders, mitochondrial genome rearrangements resulted in additional transfers of canonical genes into the P8 segment. In stony corals two versions of 6 and 11 genes are intron-located (**Figure 2A** and **B**). Furthermore, it was noted that robust-clade species have developed a highly compact ribozyme core compared to complex-clade species (and all other hexacorals) [19]. The most complex ND5-717 introns are found in mushroom corals and in the *Protanthea* sea anemone [6, 16, 28, 30]. Whereas three versions of 9, 11, and 15 intron-located genes are noted in mushroom corals, 14 genes are present in P8 of *Protanthea* (**Figure 2B** and **C**). The ND5-717 intron in mushroom corals represent the largest group I intron known to date with an approximate size of 19 kb.

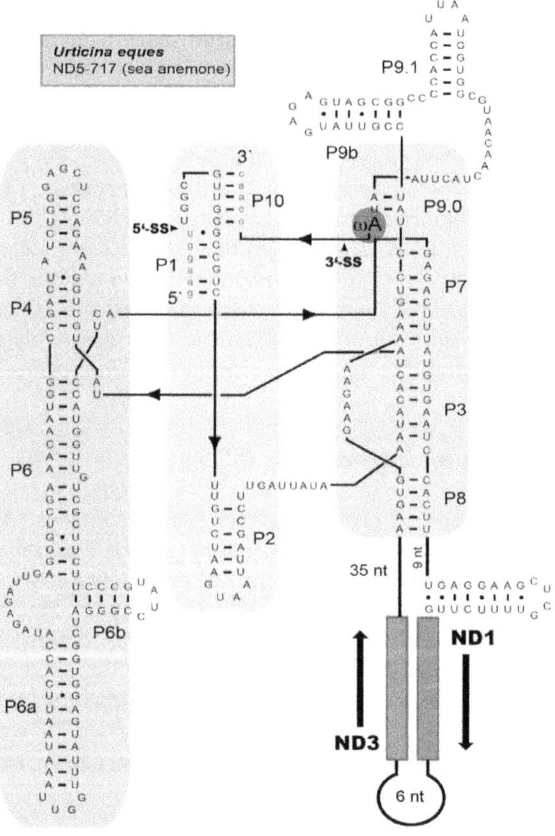

**Figure 3.**
*Structure diagram of* Urticina eques *ND5-717 group I intron. Conserved helical segments (P1–P10) are indicated, and flanking ND5 exon sequences are shown in lowercase letters. The three helical stacks, named scaffold domain, substrate domain, and catalytic domain, are indicated by blue, yellow, and green boxes, respectively. The last nucleotide of the intron (ω), which is considered as a universally conserved guanosine (ωG) in group I intron, is ωA in hexacoral ND5-717 introns (red circle). The P8 segment harbors the two OxPhos genes ND1 and ND3.*

## 3.2 Unconventional splicing of ND5-717 introns

Mitochondrial RNA sequencing reveals perfectly ligated ND5 mRNA exons in sea anemones [12, 15], colonial anemones [14], and mushroom corals [16], which support a biological splicing activity of ND5-717 introns. In the mushroom corals *Ricordea* and *Amplexodiscus*, the splicing efficiency of the ND5-717 intron was reported to be about 10% of that of the COI-884 intron located in the same mitochondrial genome [16]. The complex ND5-717 intron contains 2–15 mitochondrial genes within P8 that challenges its mode of splicing. The shortest forms of ND5-717 introns (approximately 1.6–2.4 kb) detected in sea anemones, colonial anemones, and black corals are likely to be excised by conventional group I intron *cis*-splicing from one single precursor RNA (**Figure 4A**).

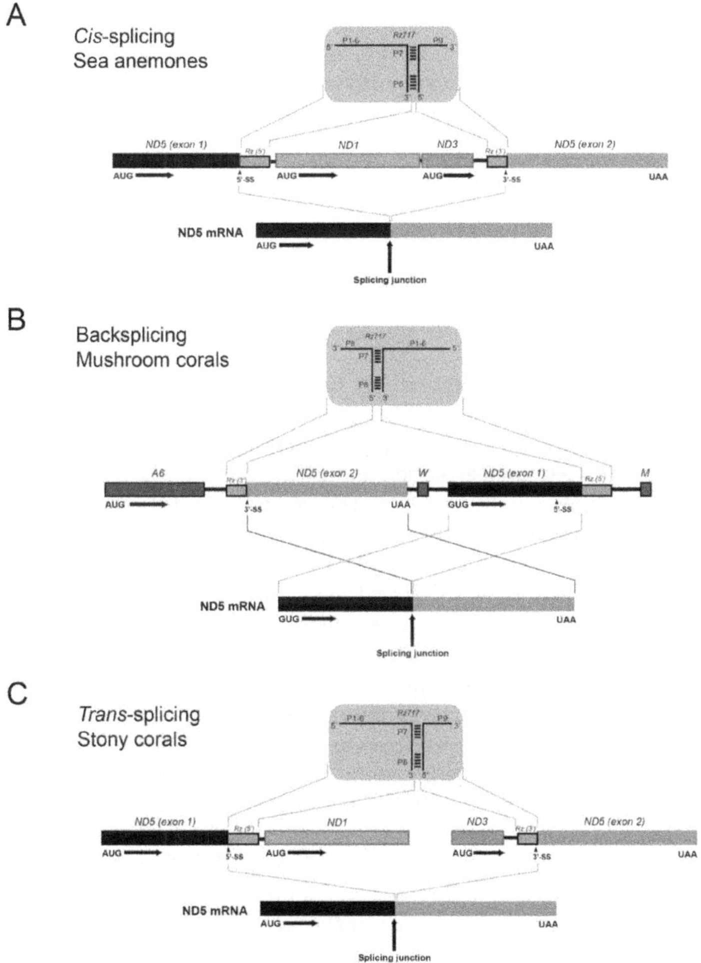

**Figure 4.**
*Different modes of ND5-717 intron splicing. A schematic group I ribozyme (Rz717; green box) is indicated above each precursor map, and ligated ND5 mRNA is shown below. Splice sites (5'SS and 3'SS), initiation codons (AUG/GUG), and stop codons (UAA) are indicated. (A) Cis-splicing performed from a single precursor RNA where both ND5 exons are in a conventional order (exon 1-exon 2). (B) Backsplicing performed from a single precursor RNA where both ND5 exons are in a non-conventional order (exon 2-exon 1). (C) Trans-splicing performed from two separate precursor RNAs, each containing one ND5 exon.*

The longest forms of ND5-717 introns (approximately 15–19 kb), present in mushroom corals [28, 30] and the deep-water *Protanthea* sea anemone [6], contain almost the entire mitochondrial genome within P8. Recently, experimental support of intron removal by backsplicing in mushroom corals was reported [16]. It was found that the primary ND5 transcript contains a permuted exon arrangement where exon 2 is followed by exon 1 (**Figure 4B**). Correct ND5 exon ligation was achieved by involving a circular exon-containing RNA intermediate, which is a hallmark of intron backsplicing [16]. This is the first example of a natural group I intron removed by backsplicing and may explain why some hexacorals tolerate giant ND5-717 group I introns.

How the ND5-717 introns in stony corals are removed from their precursors by splicing is currently not known. These introns (sizes from approximately 6–12 kb) [19, 39] may be too large and complex to be removed by conventional *cis*-splicing, and the ND5 exons may be too distant apart for backsplicing. Thus, a more plausible alternative is *trans*-splicing that generates a ligated ND5 mRNA from two separate precursor RNAs (**Figure 4C**). An interesting notion is that group I intron *trans*-splicing has been reported in mitochondrial transcripts of placozoan animals [40].

## 4. Mobile-type group I introns in the COI gene

The gene encoding COI is a frequent host of group I introns in hexacoral mitochondrial genomes. Of the total 133 species inspected (**Appendix Table 1**), about 50% harbor an intron insertion. COI introns are present in all five hexacoral orders, but at different distribution patterns.

### 4.1 Three different insertion sites in the COI gene

The COI gene is interrupted by group I introns at three genic positions, where each intron site represents a unique evolutionary history [14, 41]. The intron insertion sites correspond to positions 720, 867, and 884 (human COI gene numbering [19]). The COI-884 introns are widespread in hexacorals, present in most investigated species of sea anemones, mushroom corals, and black corals, as well as a few stony corals (**Appendix Table 1**) [12, 41, 42]. Colonial anemones harbor COI-867 introns [14], and some Indo-Pacific stony coral species contain COI-720 introns [41, 43]. It appears that hexacorals are infected at least three times by COI introns or that this mitochondrial gene is subjected to recurrent group I intron invasion and extinction.

COI introns at different insertion sites are distinct in their ribozyme secondary structure, exemplified by the *Urticina* sea anemone and *Zoanthus* colonial anemone introns COI-884 and COI-867, respectively (**Figure 5A** and **B**). A common feature, however, is the large insertion within helical segment P8 harboring a HEG that codes for a homing endonuclease of the LAGLIDADG family. These HEGs extend beyond P8 and into the ribozyme domains [12, 14, 15, 43]. Thus, COI-720, COI-867, and COI-884 intron sequences possess dual coding potentials of catalytic RNAs and homing endonucleases. This integration of the endonuclease into the ribozyme core structure ties the two elements closer together, making the endonuclease less prone to degradation.

### 4.2 Expression of intron-encoded homing endonucleases

Mobile-type introns, like the hexacoral mitochondrial COI introns, promote homing into cognate intron-less alleles by gene conversion [44, 45]. Intron homing

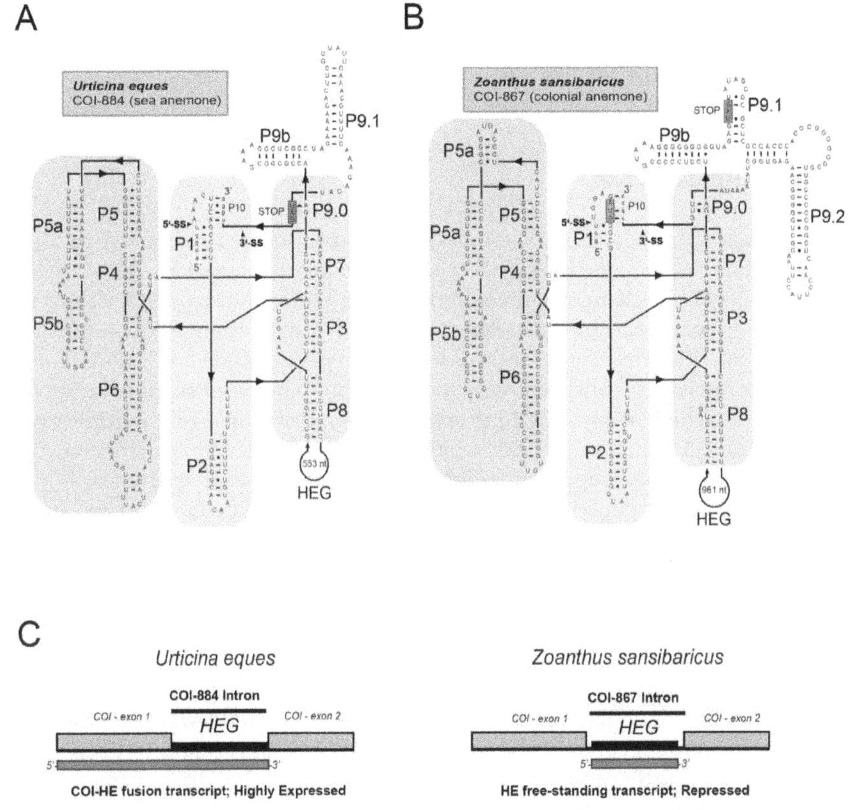


**Figure 5.**
*COI introns and HEG expression strategy. (A) Secondary structure diagram of the sea anemone* Urticina eques
*COI-884 group I intron. The conserved paired segments of the catalytic core (P1–P10) are shown, and flanking
COI exon sequences are in lowercase letters. The P8 extension containing the HEG is indicated. Note that the
HEG stop codon (UAG; red box) refers to the last three nucleotides of the intron. The three helical stacks are
indicated by blue, yellow, and green boxes. (B) Secondary structure diagram of the colonial anemone* Zoanthus
sansibaricus *COI-867 group I intron. The P1–P10 core segments are shown, and the P8 extension containing
the HEG is indicated. Flanking COI exon sequences are in lowercase letters. Note the HEG initiation codon
(AUG; green box) and stop codon (UAG; red box) are located in the 5′ end and 3′ end, respectively, of the
intron sequence. The three helical stacks are indicated by blue, yellow, and green boxes. (C) Organization of
homing endonuclease transcripts from the COI-884 intron (*Urticina eques*; left) and the COI-867 intron
(*Zoanthus sansibaricus*; right). While the COI-884 intron transcript is in-frame with the COI exon 1 and
highly expressed, the COI-867 intron transcript is freestanding within the intron and repressed. HE, homing
endonuclease.*

is initiated by a DNA double-strand break catalyzed by the intron-encoded homing
endonuclease. Expression of HEGs has been studied in COI introns of sea anemones, colonial anemones, and mushroom corals [12, 14–16]. Two main versions
were noted, leading to either highly expressed or repressed HEGs (**Figure 5C**).
(1) The most successful mode of expression is the in-frame COI-HEG fusion
strategy. The HEG, which covers most of the intron sequences (including the
ribozyme encoded parts), is fused in-frame with the 5′ COI exon. Highly expressed
in-frame HEGs are observed in the sea anemones *Urticina* and *Bolocera* [12], and
similar in-frame organizations appear common in other sea anemones such as
*Isosicyonis*, *Phymanthus*, *Actinia*, and *Stichodactyla* ([15, 46, 47]; our unpublished
results). A COI fusion strategy for intron HEG expression in mitochondria, however, is not unique to sea anemones since several fungi are using this approach
[44, 48]. (2) Truncated in-frame fusions or freestanding intron HEGs result in

significant lower expressions. This is observed for COI-884 introns of *Hormathia* and *Anemonia* sea anemones [12, 15], COI-884 introns of mushroom corals [16], and COI-867 introns of colonial anemones [14].

## 5. Concluding remarks

A hallmark of hexacoral mitochondrial genomes is the presence of self-catalytic group I introns. What is the biological role of these mitochondrial introns—are they purely selfish genetic elements, or could they have gained new regulatory functions beyond self-splicing? Current knowledge suggests a fungal origin of the hexacoral introns [19, 34, 49]. The group I introns in the COI gene encode LAGLIDADG-type homing endonucleases, consistent with intron mobility between cognate intron-less alleles [12, 45]. The hexacoral COI introns appear gained and lost in multiple cycles during the last 0.5 billion years [42], which supports a selfish intron behavior.

The ND5-717 intron is apparently obligatory in hexacoral mitochondrial genomes, making this genetic element an interesting candidate in gene regulation. Similar obligatory group I introns have been noted in the chloroplast tRNA$^{\text{Leu}}$ gene of all green plants and in the nuclear LSU rRNA gene of all Physarales myxomycetes [50, 51]. These obligatory mitochondrial, chloroplast, and nuclear introns are considered domesticated group I introns that may have gained new host-specific functions beyond self-splicing [21, 25]. The mitochondrial ND5 mRNA stability has a key role in respiratory control in higher animals; it is tightly regulated and contains m$^1$A base modification [52–54]. Intron retention of ND5 mRNA was recently reported in mushroom corals [16], suggesting possible host regulatory functions in hexacorals. Thus, further investigations on hexacoral mitochondrial intron functions and biological roles are needed and highly welcome.

## Acknowledgements

We thank current and former members of the research teams at the Genomics Group (Nord University) and the RNA Group (UiT—The Arctic University of Norway) for discussion and support. A special thanks to the former PhD students Sylvia Ighem Chi and Ilona Urbarova for thoroughly investigating hexacoral genomics.

## Conflict of interest

The authors declare that they have no conflict of interest.

## A. Appendix

In January 2020 about 200 hexacoral mitochondrial genomes have been completely, or nearly completely, sequenced. These mitochondrial genomes represent all 5 hexacoral orders, 51 families, 77 genera, and 133 distinct species. All specimens (100%) harbor ND5-717 and approximately 50% harbor COI introns. Key features are summarized in **Appendix Table 1**.

| Species | Family | Accession no | Mt size[1] | ND5 intron (size)[2] | COI intron (size)[3] |
|---|---|---|---|---|---|
| **A: Sea anemones (Order Actiniaria)** | | | | | |
| *Synhalcurias elegans* | Actiniaridae | KR051009 | P11,445 bp | ND5+ (1635 bp) | COI– |
| *Actinia equina* | Actiniidae | MH545699 | C20,690 bp | ND5+ (2170 bp) | 884_COI+ (857 bp) |
| *Actinia tenebrosa* | Actiniidae | MK291977 | C20,691 bp | ND5+ (2170 bp) | 884_COI+ (854 bp) |
| *Anemonia majano* | Actiniidae | KY860670 | C19,545 bp | ND5+ (1679 bp) | 884_COI+ (853 bp) |
| *Anemonia sulcata* | Actiniidae | MN011067 | C20,390 bp | ND5+ (1725 bp) | 884_COI+ (1053 bp) |
| *Anemonia viridis* | Actiniidae | KY860669 | C20,108 bp | ND5+ (1726 bp) | 884_COI+ (853 bp) |
| *Anthopleura midori* | Actiniidae | KT989511 | C20,039 bp | ND5+ (1714 bp) | 884_COI+ (854 bp) |
| *Bolocera tuediae* | Actiniidae | HG423145 | C19,143 bp | ND5+ (2055 bp) | 884_COI+ (853 bp) |
| *Bolocera* sp. | Actiniidae | KU507297 | C19,463 bp | ND5+ (2397 bp) | 884_COI+ (854 bp) |
| *Entacmaea quadricolor* | Actiniidae | MN066616 | C20,960 bp | ND5+ (2052 bp) | 884_COI+ (853 bp) |
| *Epiactis japonica* | Actiniidae | MN076184 | C18,835 bp | ND5+ (1681 bp) | 884_COI+ (853 bp) |
| *Epiactis prolifera* | Actiniidae | Ref. [33] | C19,752 bp | ND5+ (1737 bp) | 884_COI+ (853 bp) |
| *Isosicyonis striata* | Actiniidae | KR051006 | C19,001 bp | ND5+ (1695 bp) | 884_COI+ (853 bp) |
| *Urticina eques* | Actiniidae | HG423144 | C20,458 bp | ND5+ (1681 bp) | 884_COI+ (850 bp) |
| *Antholoba achates* | Actinostolidae | KR051002 | C17,816 bp | ND5+ (1884 bp) | 884_COI+ (853 bp) |
| *Stomphia selaginella* | Actinostolidae | Ref. [33] | C18,349 bp | ND5+ (1784 bp) | 884_COI+ (829 bp) |
| *Aiptasia pulchella*[4] | Aiptasiidae | HG423147 | C19,791 bp | ND5+ (1730 bp) | 884_COI+ (847 bp) |
| *Aiptasia pulchella*[4] | Aiptasiidae | HG423148 | C19,790 bp | ND5+ (1730 bp) | 884_COI+ (847 bp) |
| *Bartholomea annulata* | Aiptasiidae | MN066614 | C19,615 bp | ND5+ (1754 bp) | 884_COI+ (847 bp) |
| *Alicia sansibarensis* | Aliciidae | KR051001 | C19,575 bp | ND5+ (2158 bp) | COI– |
| *Relicanthus daphneae* | Boloceroididae | MK947129 | C17,727 bp | ND5+ (1721 bp) | 884_COI+ (926 bp) |

| Species | Family | Accession no | Mt size[1] | ND5 intron (size)[2] | COI intron (size)[3] |
|---|---|---|---|---|---|
| *Edwardsia gilbertensis* | Edwardsiidae | MN066615 | P17,661 bp | ND5+ (1604 bp) | COI− |
| *Edwardsia timida* | Edwardsiidae | Ref. [33] | C18,683 bp | ND5+ (1622 bp) | COI− |
| *Nematostella* sp. | Edwardsiidae | DQ643835 | C16,389 bp | ND5+ (1620 bp) | COI− |
| *Protanthea simplex* | Gonactiniidae | MH500774/75 | C21,326 bp | ND5+ (15,262 bp) | COI− |
| *Halcampoides purpurea* | Halcampoidisae | KR051003 | C18,038 bp | ND5+ (1648 bp) | 884_COI+ (856 bp) |
| *Halcurias pilatus* | Halcuriidae | KR051004 | P10,972 bp | ND5+ (1635 bp) | COI− |
| *Haloclava producta* | Haloclavidae | MN076185 | P17,416 bp | ND5+ (1681 bp) | 884_COI+ (853 bp) |
| *Hormathia digitata* | Hormathiidae | HG423146 | C18,754 bp | ND5+ (1681 bp) | 884_COI+ (853 bp) |
| *Liponema brevicorne* | Liponematidae | MN076188 | C19,143 bp | ND5+ (2055 bp) | 884_COI+ (853 bp) |
| *Metridium senile* | Metridiidae | HG423143 | C17,444 bp | ND5+ (1681 bp) | 884_COI+ (853 bp) |
| *Metridium senile* | Metridiidae | AF000023 | C17,743 bp | ND5+ (1681 bp) | 884_COI+ (853 bp) |
| *Phymanthus crucifer* | Phymanthidae | KR051007 | C19,727 bp | ND5+ (1911 bp) | 884_COI+ (865 bp) |
| *Sagartia ornata* | Sagartiidae | KR051008 | C17,446 bp | ND5+ (1671 bp) | 884_COI+ (853 bp) |
| *Heteractis aurora* | Stichodactylidae | MN076186 | C19,999 bp | ND5+ (1737 bp) | 884_COI+ (853 bp) |
| *Heteractis crispa* | Stichodactylidae | MN076187 | C18,835 bp | ND5+ (1681 bp) | 884_COI+ (853 bp) |
| *Stichodactyla helianthus* | Stichodactylidae | Ref. [33] | C19,551 bp | ND5+ (1681 bp) | 884_COI+ (866 bp) |
| *Stichodactyla helianthus* | Stichodactylidae | Unpublished[5] | C18,999 bp | ND5+ (1680 bp) | 884_COI+ (865 bp) |
| *Stichodactyla meretensii* | Stichodactylidae | Ref. [33] | C18,849 bp | ND5+ (1681 bp) | 884_COI+ (866 bp) |
| **B: Colonial anemones (Order Zoantharia)** | | | | | |
| *Savalia savaglia* | Parazoanthidae | DQ825686 | C20,764 bp | ND5+ (2052 bp) | 867_COI+ (1238 bp) |
| *Palythoa heliodiscus* | Sphenopidae | KY888673 | C20,841 bp | ND5+ (2077 bp) | 887_COI+ (1276 bp) |
| *Zoanthus sansibaricus* | Zoanthidae | KY888672 | C20,972 bp | ND5+ (2096 bp) | 867_COI+ (1327 bp) |

| Species | Family | Accession no | Mt size[1] | ND5 intron (size)[2] | COI intron (size)[3] |
|---|---|---|---|---|---|
| **C: Mushroom corals (Order Corallimorpharia)** | | | | | |
| *Corallimorphus profundus* | Corallimorphidae | KP938440 | C20,488 bp | ND5+ (12,389 bp) | 884_COI+ (1182 bp) |
| *Corynactis californica* | Corallimorphidae | KP938436 | C20,715 bp | ND5+ (10,531 bp) | 884_COI+ (1265 bp) |
| *Pseudocorynactis* sp. | Corallimorphidae | KP938437 | C21,239 bp | ND5+ (18,840 bp) | 884_COI+ (1177 bp) |
| *Amplexidiscus fenestrafer* | Discosomatidae | MH308002 | C20,054 bp | ND5+ (17,960 bp) | 884_COI+ (1206 bp) |
| *Amplexidiscus fenestrafer* | Discosomatidae | KP938435 | C20,188 bp | ND5+ (18,094 bp) | 884_COI+ (1206 bp) |
| *Discosoma nummiforme* | Discosomatidae | KP938434 | C20,925 bp | ND5+ (18,791 bp) | 884_COI+ (1208 bp) |
| *Discosoma* sp. | Discosomatidae | DQ643965 | C20,908 bp | ND5+ (18,803 bp) | 884_COI+ (1207 bp) |
| *Discosoma* sp. | Discosomatidae | DQ643966 | C20,912 bp | ND5+ (19,807 bp) | 884_COI+ (1206 bp) |
| *Discosoma* sp. | Discosomatidae | MH308003 | C20,288 bp | ND5+ (18,196 bp) | 884_COI+ (1206 bp) |
| *Rhodactis indosinensis* | Discosomatidae | KP938438 | C20,100 bp | ND5+ (18,013 bp) | 884_COI+ (1204 bp) |
| *Rhodactis mussoides* | Discosomatidae | KP938439 | C20,826 bp | ND5+ (18,721 bp) | 884_COI+ (1206 bp) |
| *Rhodactis* sp. | Discosomatidae | DQ640647 | C20,093 bp | ND5+ (18,001 bp) | 884_COI+ (1206 bp) |
| *Ricordea florida* | Ricordeidae | DQ640648 | C21,376 bp | ND5+ (19,247 bp) | 884_COI+ (1176 bp) |
| *Ricordea yuma* | Ricordeidae | MH308004 | C21,430 bp | ND5+ (19,301 bp) | 884_COI+ (1198 bp) |
| *Ricordea yuma* | Ricordeidae | MH308005 | C21,566 bp | ND5+ (19,437 bp) | 884_COI+ (1198 bp) |
| *Ricordea yuma* | Ricordeidae | KP938441 | C22,015 bp | ND5+ (19,886 bp) | 884_COI+ (1198 bp) |
| **D: Black corals (Order Antipatharia)** | | | | | |
| *Cirrhipathes lutkeni*[6] | Antipathidae | JX023266 | C20,448 bp | ND5+ (2062 bp) | 884_COI+ (1439 bp) |
| *Myriopathes japonica* | Antipathidae | JX456459 | C17,733 bp | ND5+ (1699 bp) | 884_COI+ (924 bp) |
| *Chrysopathes formosa* | Cladopathidae | DQ304771 | C18,398 bp | ND5+ (1932 bp) | COI− |

| Species | Family | Accession no | Mt size[1] | ND5 intron (size)[2] | COI intron (size)[3] |
|---|---|---|---|---|---|
| **E: Stony corals (Order Scleractinia)** | | | | | |
| *Complex clade* | | | | | |
| *Acropora aculeus* | Acroporidae | KT001202 | C18,528 bp | ND5+ (12,116 bp) | COI− |
| *Acropora acuminata* | Acroporidae | LC201815 | C18,586 bp | ND5+ (12,175 bp) | COI− |
| *Acropora aspera* | Acroporidae | KF448532 | C18,479 bp | ND5+ (12,070 bp) | COI− |
| *Acropora austera* | Acroporidae | LC201816 | C18,346 bp | ND5+ (11,937 bp) | COI− |
| *Acropora awi* | Acroporidae | LC201849 | C18,478 bp | ND5+ (12,070 bp) | COI− |
| *Acropora awi* | Acroporidae | LC201850 | C18,479 bp | ND5+ (12,070 bp) | COI− |
| *Acropora awi* | Acroporidae | LC201851 | C18,479 bp | ND5+ (12,070 bp) | COI− |
| *Acropora awi* | Acroporidae | LC201852 | C18,479 bp | ND5+ (12,070 bp) | COI− |
| *Acropora awi* | Acroporidae | LC201853 | C18,479 bp | ND5+ (12,070 bp) | COI− |
| *Acropora awi* | Acroporidae | LC201854 | C18,479 bp | ND5+ (12,070 bp) | COI− |
| *Acropora awi* | Acroporidae | LC201855 | C18,479 bp | ND5+ (12,070 bp) | COI− |
| *Acropora carduus* | Acroporidae | LC201813 | C18,373 bp | ND5+ (11,964 bp) | COI− |
| *Acropora carduus* | Acroporidae | LC201814 | C18,372 bp | ND5+ (11,963 bp) | COI− |
| *Acropora cytherea* | Acroporidae | LC201817 | C18,568 bp | ND5+ (12,158 bp) | COI− |
| *Acropora cytherea* | Acroporidae | LC201818 | C18,567 bp | ND5+ (12,157 bp) | COI− |
| *Acropora cytherea* | Acroporidae | LC201819 | C18,568 bp | ND5+ (12,158 bp) | COI− |
| *Acropora digitifera* | Acroporidae | KF448535 | C18,479 bp | ND5+ (12,070 bp) | COI− |
| *Acropora divaricata* | Acroporidae | KF448537 | C18,481 bp | ND5+ (12,072 bp) | COI− |
| *Acropora echinata* | Acroporidae | LC201820 | C18,480 bp | ND5+ (12,071 bp) | COI− |
| *Acropora echinata* | Acroporidae | LC201821 | C18,480 bp | ND5+ (12,071 bp) | COI− |

| Species | Family | Accession no | Mt size[1] | ND5 intron (size)[2] | COI intron (size)[3] |
|---|---|---|---|---|---|
| *Acropora echinata* | Acroporidae | LC201822 | C18,480 bp | ND5+ (12,071 bp) | COI− |
| *Acropora echinata* | Acroporidae | LC201823 | C18,480 bp | ND5+ (12,071 bp) | COI− |
| *Acropora echinata* | Acroporidae | LC201824 | C18,480 bp | ND5+ (12,071 bp) | COI− |
| *Acropora echinata* | Acroporidae | LC201825 | C18,480 bp | ND5+ (12,071 bp) | COI− |
| *Acropora echinata* | Acroporidae | LC201826 | C18,482 bp | ND5+ (12,071 bp) | COI− |
| *Acropora echinata* | Acroporidae | LC201834 | C18,481 bp | ND5+ (12,072 bp) | COI− |
| *Acropora echinata* | Acroporidae | LC201835 | C18,368 bp | ND5+ (11,959 bp) | COI− |
| *Acropora echinata* | Acroporidae | LC201836 | C18,482 bp | ND5+ (12,071 bp) | COI− |
| *Acropora echinata* | Acroporidae | LC201837 | C18,368 bp | ND5+ (11, 959 bp) | COI− |
| *Acropora echinata* | Acroporidae | LC201838 | C18,480 bp | ND5+ (12,071 bp) | COI− |
| *Acropora echinata* | Acroporidae | LC201839 | C18,482 bp | ND5+ (12,073 bp) | COI− |
| *Acropora echinata* | Acroporidae | LC201840 | C18,480 bp | ND5+ (12,071 bp) | COI− |
| *Acropora echinata* | Acroporidae | LC201841 | C18,367 bp | ND5+ (11,958 bp) | COI− |
| *Acropora florida* | Acroporidae | KF448533 | C18,365 bp | ND5+ (11,956 bp) | COI− |
| *Acropora florida* | Acroporidae | LC201827 | C18,365 bp | ND5+ (11,956 bp) | COI− |
| *Acropora grandis* | Acroporidae | LC201828 | C18,479 bp | ND5+ (12,070 bp) | COI− |
| *Acropora horrida* | Acroporidae | KF448530 | C18,480 bp | ND5+ (12,071 bp) | COI− |
| *Acropora humilis* | Acroporidae | KF448528 | C18,479 bp | ND5+ (12,070 bp) | COI− |
| *Acropora hyacinthus* | Acroporidae | KF448531 | C18,566 bp | ND5+ (12,157 bp) | COI− |
| *Acropora hyacinthus* | Acroporidae | LC201829 | C18,567 bp | ND5+ (12,157 bp) | COI− |
| *Acropora hyacinthus* | Acroporidae | LC201830 | C18,567 bp | ND5+ (12,157 bp) | COI− |
| *Acropora hyacinthus* | Acroporidae | LC201831 | C18,567 bp | ND5+ (12,157 bp) | COI− |

| Species | Family | Accession no | Mt size[1] | ND5 intron (size)[2] | COI intron (size)[3] |
|---|---|---|---|---|---|
| *Acropora hyacinthus* | Acroporidae | LC201832 | C18,568 bp | ND5+ (12,158 bp) | COI– |
| *Acropora intermedia* | Acroporidae | LC201833 | C18,479 bp | ND5+ (12,070 bp) | COI– |
| *Acropora microphthalma* | Acroporidae | LC201842 | C18,479 bp | ND5+ (12,070 bp) | COI– |
| *Acropora microphthalma* | Acroporidae | LC201843 | C18,481 bp | ND5+ (12,072 bp) | COI– |
| *Acropora muricata* | Acroporidae | KF448529 | C18,481 bp | ND5+ (12,072 bp) | COI– |
| *Acropora muricata* | Acroporidae | LC201844 | C18,480 bp | ND5+ (12,071 bp) | COI– |
| *Acropora nasuta* | Acroporidae | KF448536 | C18,481 bp | ND5+ (12,072 bp) | COI– |
| *Acropora nasuta* | Acroporidae | LC201845 | C18,374 bp | ND5+ (11,965 bp) | COI– |
| *Acropora nasuta* | Acroporidae | LC201846 | C18,484 bp | ND5+ (12,074 bp) | COI– |
| *Acropora robusta* | Acroporidae | KF448538 | C18,480 bp | ND5+ (12,071 bp) | COI– |
| *Acropora selago* | Acroporidae | LC201847 | C18,482 bp | ND5+ (12,073 bp) | COI– |
| *Acropora selago* | Acroporidae | LC201848 | C18,480 bp | ND5+ (12,071 bp) | COI– |
| *Acropora tenuis* | Acroporidae | AF338425 | C18,338 bp | ND5+ (11,928 bp) | COI– |
| *Acropora tenuis* | Acroporidae | LC201856 | C18,342 bp | ND5+ (11,933 bp) | COI– |
| *Acropora tenuis* | Acroporidae | LC201857 | C18,342 bp | ND5+ (11,933 bp) | COI– |
| *Acropora tenuis* | Acroporidae | LC201858 | C18,343 bp | ND5+ (11,934 bp) | COI– |
| *Acropora tenuis* | Acroporidae | LC201859 | C18,342 bp | ND5+ (11,933 bp) | COI– |
| *Acropora tenuis* | Acroporidae | LC201860 | C18,342 bp | ND5+ (11,933 bp) | COI– |
| *Acropora tenuis* | Acroporidae | LC201861 | C18,342 bp | ND5+ (11,933 bp) | COI– |
| *Acropora tenuis* | Acroporidae | LC201862 | C18,343 bp | ND5+ (11,934 bp) | COI– |
| *Acropora tenuis* | Acroporidae | LC201863 | C18,342 bp | ND5+ (11,933 bp) | COI– |
| *Acropora tenuis* | Acroporidae | LC201864 | C18,342 bp | ND5+ (11,933 bp) | COI– |

| Species | Family | Accession no | Mt size[1] | ND5 intron (size)[2] | COI intron (size)[3] |
|---|---|---|---|---|---|
| *Acropora tenuis* | Acroporidae | LC201865 | C18,342 bp | ND5+ (11,933 bp) | COI− |
| *Acropora tenuis* | Acroporidae | LC201866 | C18,342 bp | ND5+ (11,933 bp) | COI− |
| *Acropora tenuis* | Acroporidae | LC201867 | C18,342 bp | ND5+ (11,933 bp) | COI− |
| *Acropora tenuis* | Acroporidae | LC201868 | C18,342 bp | ND5+ (11,933 bp) | COI− |
| *Acropora tenuis* | Acroporidae | LC201869 | C18,341 bp | ND5+ (11,933 bp) | COI− |
| *Acropora tenuis* | Acroporidae | LC201870 | C18,342 bp | ND5+ (11,933 bp) | COI− |
| *Acropora valida* | Acroporidae | MH141598 | C18,385 bp | ND5+ (11,976 bp) | COI− |
| *Acropora yongei* | Acroporidae | KF448534 | C18,342 bp | ND5+ (11,933 bp) | COI− |
| *Anacropora matthai* | Acroporidae | AY903295 | C17,888 bp | ND5+ (11,492 bp) | COI− |
| *Astreopora explanata* | Acroporidae | KJ634269 | C18,106 bp | ND5+ (11,795 bp) | COI− |
| *Astreopora myriophthalma* | Acroporidae | KJ634272 | C18,106 bp | ND5+ (11,795 bp) | COI− |
| *Montipora cactus* | Acroporidae | AY903296 | C17,887 bp | ND5+ (11,485 bp) | COI− |
| *Montipora aequituberculata* | Acroporidae | KU762339 | C17,886 bp | ND5+ (11,488 bp) | COI− |
| *Montipora efflorescens* | Acroporidae | MG851914 | C17,886 bp | ND5+ (11,491 bp) | COI− |
| *Agaricia fragilis* | Agariciidae | KM051016 | C18,667 bp | ND5+ (11,525 bp) | COI− |
| *Agaricia humilis* | Agariciidae | DQ643831 | C18,735 bp | ND5+ (11,536 bp) | COI− |
| *Pavona clavus* | Agariciidae | DQ643836 | C18,315 bp | ND5+ (11,129 bp) | COI− |
| *Pavona decussata* | Agariciidae | KP231535 | C18,378 bp | ND5+ (11,129 bp) | COI− |
| *Dendrophyllia arbuscula* | Dendrophyllidae | KR824937 | C19,069 bp | ND5+ (11,299 bp) | 884_COI+ (964 bp) |
| *Dendrophyllia cribrosa* | Dendrophyllidae | JQ290080 | C19,072 bp | ND5+ (11,282 bp) | 884_COI+ (964 bp) |
| *Tubastraea coccinea* | Dendrophyllidae | KX024566 | C19,094 bp | ND5+ (11,322 bp) | 884_COI+ (964 bp) |
| *Tubastraea coccinea* | Dendrophyllidae | JQ290078 | C19,070 bp | ND5+ (11,300 bp) | 884_COI+ (964 bp) |

| Species | Family | Accession no | Mt size[1] | ND5 intron (size)[2] | COI intron (size)[3] |
|---|---|---|---|---|---|
| *Tubastraea tagusensis* | Dendrophylliidae | KX024567 | C19,094 bp | ND5+ (11,324 bp) | 884_COI+ (964 bp) |
| *Turbinaria peltata* | Dendrophylliidae | KJ725201 | C18,966 bp | ND5+ (11,332 bp) | 884_COI+ (964 bp) |
| *Euphyllia ancora* | Euphylliidae | JF825139 | C18,875 bp | ND5+ (11,866 bp) | COI− |
| *Galaxea fascicularis* | Euphylliidae | KU159433 | C18,751 bp | ND5+ (12,022 bp) | COI− |
| *Fungiacyathus stephanus* | Fungiacyathidae | JF825138 | C19,381 bp | ND5+ (10,932 bp) | COI+ (961 bp) |
| *Alveopora japonica* | Poritidae | MG851913 | C18,144 bp | ND5+ (11,621 bp) | COI− |
| *Alveopora* sp. | Poritidae | KJ634271 | C18,146 bp | ND5+ (11,621 bp) | COI− |
| *Goniopora columna* | Poritidae | JF825141 | C18,766 bp | ND5+ (11,175 bp) | 884_COI+ (964 bp) |
| *Porites fontanesii* | Poritidae | NC_037434 | C18,658 bp | ND5+ (11,131 bp) | 884_COI+ (965 bp) |
| *Porites harrisoni* | Poritidae | NC_037435 | C18,630 bp | ND5+ (11,133 bp) | 884_COI+ (965 bp) |
| *Porites lobata* | Poritidae | KU572435 | C18,647 bp | ND5+ (11,133 bp) | 884_COI+ (965 bp) |
| *Porites lutea* | Poritidae | KU159432 | C18,646 bp | ND5+ (11,130 bp) | 884_COI+ (971 bp) |
| *Porites okinawensis* | Poritidae | JF825142 | C18,647 bp | ND5+ (11,133 bp) | 884_COI+ (965 bp) |
| *Porites panamensis* | Poritidae | KJ546638 | C18,628 bp | ND5+ (11,117 bp) | 884_COI+ (965 bp) |
| *Porites poritis* | Poritidae | DQ643837 | C18,648 bp | ND5+ (11,135 bp) | 884_COI+ (965 bp) |
| *Porites rus* | Poritidae | LN864762 | C18,647 bp | ND5+ (11,133 bp) | 884_COI+ (971 bp) |
| *Pseudosiderastrea formosa* | Siderastreidae | KP260632 | C19,475 bp | ND5+ (11,524 bp) | 884_COI+ (970 bp) |
| *Pseudosiderastrea tayami* | Siderastreidae | KP260633 | C19,475 bp | ND5+ (11,524 bp) | 884_COI+ (970 bp) |
| *Siderastrea radians* | Siderastreidae | DQ643838 | C19,387 bp | ND5+ (11,463 bp) | 884_COI+ (988 bp) |
| *Robust clade* | | | | | |
| *Madracis decactis* | Astrocoeniidae | KX982259 | C16,970 bp | ND5+ (10,435 bp) | COI− |
| *Madracis mirabilis* | Astrocoeniidae | EU400212 | C16,951 bp | ND5+ (10,415 bp) | COI− |

| Species | Family | Accession no | Mt size[1] | ND5 intron (size)[2] | COI intron (size)[3] |
|---|---|---|---|---|---|
| *Lopheila pertusa*[7] | Caryophyllidae | FR821799 | C16,150 bp | ND5+ (6460 bp) | COI– |
| *Lophelia pertusa*[7] | Caryophyllidae | KC875348 | C16,149 bp | ND5+ (6460 bp) | COI– |
| *Lophelia pertusa*[7] | Caryophyllidae | KC875349 | C16,149 bp | ND5+ (6460 bp) | COI– |
| *Solenosmilia variabilis* | Caryophyllidae | KM609293 | C15,968 bp | ND5+ (6459 bp) | COI– |
| *Solenosmilia variabilis* | Caryophyllidae | KM609294 | C15,968 bp | ND5+ (6459 bp) | COI– |
| *Colpopyllia natans* | Flaviidae | DQ643833 | C16,906 bp | ND5+ (10,445 bp) | COI– |
| *Plesiastrea versipora* | Flaviidae | MH025639 | C15,320 bp | ND5+ (9398 bp) | COI– |
| *Echinophyllia aspera* | Lobophylliidae | MG792550 | C17,697 bp | ND5+ (10,136 bp) | 720_COI+ (1077 bp) |
| *Sclerophyllia maxima*[8] | Lobophylliidae | FO904931 | C18,168 bp | ND5+ (10,760 bp) | 720_COI+ (1074 bp) |
| *Dipsastraea rotumana* | Merulinidae | MH119077 | C16,466 bp | ND5+ (10,149 bp) | COI– |
| *Flavites halicora* | Merulinidae | MH794283 | C17,033 bp | ND5+ (11,150 bp) | COI– |
| *Hydnopora exesa* | Merulinidae | MH086217 | C17,790 bp | ND5+ (10,243 bp) | COI– |
| *Orbicella annularis* | Merulinidae | AP008973 | C16,138 bp | ND5+ (9540 bp) | COI– |
| *Orbicella annularis* | Merulinidae | AP008974 | C16,138 bp | ND5+ (9540 bp) | COI– |
| *Orbicella faveolata* | Merulinidae | AP008977 | C16,138 bp | ND5+ (9540 bp) | COI– |
| *Orbicella faveolata* | Merulinidae | AP008978 | C16,138 bp | ND5+ (9540 bp) | COI– |
| *Orbicella franksi* | Merulinidae | AP008975 | C16,138 bp | ND5+ (9540 bp) | COI– |
| *Orbicella franksi* | Merulinidae | AP008976 | C16,137 bp | ND5+ (9539 bp) | COI– |
| *Polycyathus* sp. | Merulinidae | JF825140 | C15,357 bp | ND5+ (9438 bp) | COI– |
| *Platygyra carnosa* | Merulinidae | JX911333 | C16,463 bp | ND5+ (10,164 bp) | COI– |
| *Mussa angulosa* | Mussidae | DQ643834 | C17,245 bp | ND5+ (10,636 bp) | COI– |
| *Madrepora oculata* | Oculinidae | JX236041 | C15,841 bp | ND5+ (10,140 bp) | COI– |

| Species | Family | Accession no | Mt size[1] | ND5 intron (size)[2] | COI intron (size)[3] |
|---|---|---|---|---|---|
| *Pocillopora damicornis* | Pocilloporidae | EU400213 | C17,425 bp | ND5+ (10,864 bp) | COI– |
| *Pocillopora damicornis* | Pocilloporidae | EF526302 | C17,415 bp | ND5+ (10,863 bp) | COI– |
| *Pocillopora eydouxi* | Pocilloporidae | EF526303 | C17,422 bp | ND5+ (10,863 bp) | COI– |
| *Seriatopora caliendrum* | Pocilloporidae | EF633601 | C17,010 bp | ND5+ (10,467 bp) | COI– |
| *Seriatopora hystrix* | Pocilloporidae | EF633600 | C17,059 bp | ND5+ (10,465 bp) | COI– |
| *Stylophora pistillata* | Pocilloporidae | EU400214 | C17,177 bp | ND5+ (10,583 bp) | COI– |
| *Astrangia* sp. | Rhizangiidae | DQ643832 | C14,853 bp | ND5+ (9258 bp) | COI– |

[1] *Size of mitochondrial genome. C, completely sequenced; P, partial/almost completely sequenced.*
[2] *Size of ND5-717 group I intron.*
[3] *Size of COI group I intron. COI–, no COI intron present; 720, 867, or 884 introns indicated.*
[4] *The sea anemone Aiptasia pulcella may also be annotated as Exaiptasia pallida.*
[5] *Information from our unpublished complete mitochondrial genome sequence of Stichodactyla helianthus.*
[6] *The black coral Cirrhipathes lutkeni may also be annotated as Strichpates lutkeni.*
[7] *The stony coral Lophelia pertusa may also be annotated as Desmophyllum pertusum.*
[8] *The stony coral Sclerophyllia maxima may also be annotated as Acanthastrea maxima.*

**Appendix Table 1.**
*Key features of group I introns in hexacoral mitogenomes.*

## Author details

Steinar Daae Johansen[1*] and Åse Emblem[2]

1 Genomics Group, Faculty of Biosciences and Aquaculture, Nord University, Bodø, Norway

2 Research Laboratory and Department of Laboratory Medicine, Nordland Hospital, Bodø, Norway

*Address all correspondence to: steinar.d.johansen@nord.no

IntechOpen

# References

[1] Daly M, Brugler MR, Cartwright P, Collins AG, Dawson MN, Fautin DG, et al. The phylum Cnidaria: A review of phylogenetic patterns and diversity 300 years after Linnaeus. Zootaxa. 1668; **2007**:127-182. Available from: www.mapress.com/zootaxa/

[2] Stampar SN, Maronna MM, Kitahara MV, Reimer JD, Morandini AC. Fast-evolving mitochondrial DNA in Ceriantharia: A reflection of hexacorallia paraphyly? PLoS One. 2014;**9**:e86612. DOI: 10.1371/journal.pone.0086612

[3] Roberts JM, Wheeler AJ, Freiwald A. Reefs of the deep: The biology and geology of cold-water coral ecosystems. Science. 2006;**312**:543-547. DOI: 10.1126/science.1119861

[4] Daly M. *Boloceroides daphneae*, a new species of giant sea anemone (Cnidaria: Actiniaria: Boloceroididae) from the deep Pacific. Marine Biology. 2006;**148**: 1241-1247. DOI: 10.1007/s00227-005-0170-7

[5] Zhang B, Zhang Y-H, Wang X, Zhang H-X, Lin Q. The mitochondrial genome of a sea anemone *Bolocera* sp. exhibits novel genetic structures potentially involved in adaptation to the deep-sea environment. Ecology and Evolution. 2017;7:4951-4962. DOI: 10.1002/ece3.3067

[6] Dubin A, Chi SI, Emblem Å, Moum T, Johansen SD. Deep-water sea anemone with a two-chromosome mitochondrial genome. Gene. 2019;**692**:195-200. DOI: 10.1016/j.gene.2018.12.074

[7] Roberts JM, Wheeler AJ, Freiwald A, Cairns S. Cold-water Corals: The Biology and Geology of Deep-sea Coral Habitats. New York: Cambridge University Press; 2009. ISBN: 978-0-521-88485-3

[8] Buhl-Mortensen L, Buhl-Mortensen P. Cold Temperature Coral Habitats, Corals in a Changing World, CD Beltran and ET Camacho. Rijeka: IntechOpen; 2018. DOI: 10.5772/intechopen.71446

[9] Friedman JR, Nunnari J. Mitochondrial form and function. Nature. 2014;**505**:335-343. DOI: 10.1038/nature12985

[10] Anderson AJ, Jackson TD, Stroud DA, Stojanovski D. Mitochondria—Hubs for regulating cellular biochemistry: Emerging concepts and networks. Open Biology. 2019;**9**: 190126. DOI: 10.1098/rsob.190126

[11] Osigus HJ, Eitel M, Bernt M, Donath A, Schierwater B. Mitogenomics at the base of Metazoa. Molecular Phylogenetics and Evolution. 2013;**69**: 339-351. DOI: 10.16/j.ympev.2013.07.016

[12] Emblem Å, Okkenhaug S, Weiss ES, Denver DR, Karlsen BO, Moum T, et al. Sea anemones possess dynamic mitogenome structures. Molecular Phylogenetics and Evolution. 2014;**75**: 184-193. DOI: 10.1016/j.ympev.2014.02.016

[13] Flot JF, Tillier S. The mitochondrial genome of *Pocillopora* (Cnidaria: Scleractinia) contains two variable regions: The putative D-loop and a novel ORF of unknown function. Gene. 2007;**401**:80-87. DOI: 10.1016/j.gene.2007.07.006

[14] Chi SI, Johansen SD. Zoantharian mitochondrial genomes contain unique complex group I introns and highly conserved intergenic regions. Gene. 2017;**628**:24-31. DOI: 10/1016/j.gene.2017.07.023

[15] Chi SI, Urbarova I, Johansen SD. Expression of homing endonuclease gene and insertion-like element in sea anemone mitochondrial genomes: Lesson learned from *Anemonia viridis*. Gene. 2018;**652**:78-86. DOI: 10.1016/j.gene.2018.01.067

[16] Chi SI, Dahl M, Emblem Å, Johansen SD. Giant group I intron in a mitochondrial genome is removed by RNA back-splicing. BMC Molecular Biology. 2019;**20**:16. DOI: 10.1186/s12867-019-0134-y

[17] Beagley CT, Okada NA, Wolstenholme DR. Two mitochondrial group I introns in a metazoan, the sea anemone *Metridium senile*: One intron contains genes for subunits 1 and 3 of NADH dehydrogenase. Proceedings of the National Academy of Sciences of the United States of America. 1996;**93**:5619-5623. DOI: 10.1073/pnas.93.11.5619

[18] Johansen SD, Emblem Å, Karlsen BO, Okkenhaug S, Hansen H, Moum T, et al. Approaching marine bioprospecting in hexacorals by RNA deep sequencing. New Biotechnology. 2010;**27**:267-275. DOI: 10.1016/j.nbt.2010.02.019

[19] Emblem Å, Karlsen BO, Evertsen J, Johansen SD. Mitogenome rearrangement in the cold-water scleractinian coral *Lophelia pertusa* (Cnidaria, Anthozoa) involves a long-term evolving group I intron. Molecular Phylogenetics and Evolution. 2011;**61**:495-503. DOI: 10.1016/j.ympev.2011.07.012

[20] Beagley CT, Wolstenholme DR. Characterization and localization of mitochondrial DNA-encoded tRNA and nuclear DNA-encoded tRNAs in the sea anemone *Metridium senile*. Current Genetics. 2013;**59**:139-152. DOI: 10.1007/s00294-013-0395-9

[21] Nielsen H, Johansen SD. Group I introns: Moving in new directions. RNA Biology. 2009;**6**:375-383. DOI: 10.4161/rna.6.4.9334

[22] Schuster A, Lopez JV, Becking LE, Kelly M, Pomponi SA, Worheide G, et al. Evolution of group I introns in Porifera: New evidence for intron mobility and implications for DNA barcoding. BMC Evolutionary Biology. 2017;**17**:82. DOI: 10.1186/s12862-017-0928-9

[23] Cech TR, Damberger SH, Gutell RR. Representation of the secondary and tertiary structure of group I introns. Nature Structural Biology. 1994;**1**:273-280. DOI: 10.1038/nsb0594-273

[24] Vicens Q, Cech TR. Atomic level architecture of group I introns revealed. Trends in Biochemical Sciences. 2006;**31**:41-51. DOI: 10.1016/j.tibs.2005.11.008

[25] Hedberg A, Johansen SD. Nuclear group I introns in self-splicing and beyond. Mobile DNA. 2013;**4**:17. DOI: 10.1186/1759-8753-4-17

[26] Jørgensen TE, Johansen SD. Expanding the coding potential of vertebrate mitochondrial genomes: Lesson learned from the Atlantic cod. In: Seligmann H, editor. Mitochondrial DNA—New Insight. Rijeka: IntechOpen; 2018. DOI: 10.5772/intechopen.75883

[27] Kühlbrandt W. Structure and function of mitochondrial membrane protein complexes. BMC Biology. 2015;**13**:89. DOI: 10.1186/s12915-015-0201-x

[28] Medina M, Collins AG, Takaoka TL, Kuehl JV, Boore JL. Naked corals: Skeleton loss in Scleractinia. Proceedings of the National Academy of Sciences of the United States of America. 2006;**103**:9096-9100. DOI: 10.1073/pnas.0602444103

[29] Sinniger F, Chevaldonne P, Pawlowski J. Mitochondrial genome of *Savalia savaglia* (Cnidaria, Hexacorallia) and early metazoan phylogeny. Journal of Molecular Evolution. 2007;**64**:196-203. DOI: 10.1007/s00239-006-0015-0

[30] Lin MF, Kitahara MV, Luo H, Tracey D, Geller J, Fukami H, et al.

Mitochondrial genome rearrangements in the scleractinia/corallimorpharia complex: Implications for coral phylogeny. Genome Biology and Evolution. 2014;**6**:1086-1095. DOI: 10.1093/gbe/evu084

[31] Ott M, Amunts A, Brown A. Organization and regulation of mitochondrial protein synthesis. Annual Review of Biochemistry. 2016;**85**:77-101. DOI: 10.1146/annurev-biochem-060815-014334

[32] Lin MF, Kitahara MV, Tachikawa H, Fukami H, Miller DJ, Chen CA. Novel organization of the mitochondrial genome in the deep-sea coral, *Madrepora oculata* (Hexacorallia, Scleractinia, Oculinidae) and its taxonomic implications. Molecular Phylogenetics and Evolution. 2012;**65**: 323-328. DOI: 10.1016/j. ympev.2012.06.011

[33] Xiao M, Brugler MR, Broe MB, Gusmão LC, Daly M, Rodríguez E. Mitogenomics suggests a sister relationship of *Relicanthus daphneae* (Cnidaria: Anthozoa: Hexacorallia: incerti ordinis) with Actiniaria. Scientific Reports. 2019;**9**:18182. DOI: 10.1038/s41598-019-54637-6

[34] Zubaer A, Wai A, Hausner G. The fungal mitochondrial Nad5 pan-genic intron landscape. Mitochondrial DNA Part A. 2019;**30**:835-842. DOI: 10.1080/ 24701394.2019.1687691

[35] Nelson MA, Macino G. Three class I introns in the ND4L/ND5 transcriptional unit of *Neurospora crassa* mitochondria. Molecular and General Genetics. 1987;**206**:318-325. DOI: 10.1007/bf00333590

[36] Kerscher S, Durstewitz G, Casaregola S, Gaillardin C, Brandt U. The complete mitochondrial genome of *Yarrowia lipolytica*. Comparative Functional Genomics. 2001;**2**:80-90. DOI: 10.1002/cfg.72

[37] Burger G, Forget L, Zhu Y, Gray WW, Lang BF. Unique mitochondrial genome architecture in unicellular relatives of animals. Proceedings of the National Academy of Sciences of the United States of America. 2003;**100**:892-897. DOI: 10.1073/pnas.0336115100

[38] Nielsen H, Fiskaa T, Birgisdottir AB, Haugen P, Einvik C, Johansen SD. The ability to form full-length intron RNA circles is a general property of nuclear group I introns. RNA. 2003;**9**: 1464-1475. DOI: 10.1261/rna.5290903

[39] van Oppen MJH, Catmull J, McDonald BJ, Hisop NR, Hagerman PJ, Miller DJ. The mitochondrial genome of *Acropora tenuis* (Cnidaria; Scleractinia) contains a large group I intron and a candidate control region. Journal of Molecular Evolution. 2002;**55**:1-13. DOI: 10.1007/s00239-001-0075-0

[40] Burger G, Yan Y, Javadi P, Lang FB. Group I-intron trans-splicing and mRNA editing in the mitochondria of placozoan animals. Trends in Genetics. 2009;**25**:381-386. DOI: 10.1016/j. tig.2009-07.003

[41] Celis JS, Edgell DR, Stelbrink B, Wibberg D, Hauffe T, Blom J, et al. Evolutionary and biogeographical implications of degraded LAGLIDADG endonuclease functionality and group I intron occurrence in stony corals (Scleractinia) and mushroom corals (Corallimorpharia). PLoS One. 2017;**12**: e0173734. DOI: 10.1371/journal. pone.0173734

[42] Goddard MR, Leigh J, Roger AJ, Pemberton AJ. Invasion and persistence of a selfish gene in the Cnidaria. PLoS One. 2006;**1**:e3. DOI: 10.1371/journal. pone.0000003

[43] Fukami H, Chen CA, Chiou CY, Knowlton N. Novel group I introns encoding a putative homing endonuclease in the mitochondrial cox1

gene of Scleractinian corals. Journal of Molecular Evolution. 2007;**64**:591-6009. DOI: 10.1007/s00239-006-0279-4

[44] Lambowitz AM, Belfort M. Introns as mobile genetic elements. Annual Review of Biochemistry. 1993;**62**: 587-622. DOI: 10.1146/annurev.bi.62. 070193.003103

[45] Haugen P, Simon DM, Bhattacharya D. The natural history of group I introns. Trends in Genetics. 2005;**21**:111-119. DOI: 10.1016/j.tig. 2004.12.007

[46] Foox J, Brugler M, Siddall EM, Rodriguez E. Multiplexed pyrosequencing of nine sea anemone (Cnidaria: Anthozoa: Hexacorallia: Actiniaria) mitochondrial genomes. Mitochondrial DNA Part A. 2016;**27**:2826-2832. DOI: 10.3109/ 19401736.2015.1053114

[47] Wilding CS, Weedall GD. Morphotypes of the common beadlet anemone *Actinia equina* (L.) are genetically distinct. Journal of Experimental Marine Biology and Ecology. 2019;**510**:81-85. DOI: 10.1016/j. jemb.2018.10.001

[48] Guo WW, Moran JV, Hoffman PW, Henke RM, Butow RA, Perlman PS. The mobile group I intron 3α of the yeast mitochondrial COXI gene encodes a 35-kDa processed protein that is an endonuclease but not a maturase. Journal of Biological Chemistry. 1995; **270**:15563-15570. DOI: 10.1074/jbc. 270.26.15563

[49] Férandon C, Moukha S, Callac P, Benedetto J-P, Castroviejo M, Barroso G. The *Agaricus bisporus* cox1 gene: The longest mitochondrial gene and the largest reservoir of mitochondrial group I introns. PLoS One. 2010;**5**:e14048. DOI: 10.1371/ journal.pone.0014048

[50] Kuhsel MG, Strickland R, Palmer JD. An ancient group I intron shared by eubacteria and chloroplasts. Science. 1990;**250**:1570-1573. DOI: 10.1126/science.2125748

[51] Wikmark OG, Haugen P, Haugli K, Johansen SD. Obligatory group I introns with unusual features at positions 1949 and 2449 in nuclear LSU rDNA of Didymiaceae myxomycetes. Molecular Phylogenetics and Evolution. 2007;**43**: 596-604. DOI: 10.16/j.ympev.2006. 11.004

[52] Bai Y, Shakeley RM, Attardi G. Tight control of respiration by NADH dehydrogenase ND5 subunit gene expression in mouse mitochondria. Molecular and Cellular Biology. 2000; **20**:805-815. DOI: 10.1128/ mcb.20.3.805-815.2000

[53] Chomyn A. Mitochondrial genetic control of assembly and function of complex I in mammalian cells. Journal of Bioenergetics and Biomembranes. 2001;**33**:251-257. DOI: 10.1023/a: 1010791204961

[54] Safra M, Sas-Chen A, Nir R, Winkler R, Nachshon A, Bar-Yaacov D, et al. The m$^1$A landscape on cytosolic and mitochondrial mRNA at single-base resolution. Nature. 2017;**551**:251-255. DOI: 10.1038/nature24456

Lightning Source UK Ltd.
Milton Keynes UK
UKHW030923090522
402620UK00001B/26